Novice and Technican No-Code *Plus*

Elements 2 and 3A
Novice and Technician Class Theory

BY
GORDON WEST
WB6NOA

FIRST EDITION

PROMPT® Publications is an imprint of Howard W. Sams & Company,
2647 Waterfront Parkway, East Drive, Indianapolis, IN 46214-2041.

This book was originally developed and published as <u>*No-Code Plus*</u> *by:*
  Master Publishing, Inc.
  14 Canyon Creek Village MS31
  Richardson, Texas 75080
  (214) 907-8938

International Standard Book Number: 0-7906-1047-7

Edited by: *Gerald Luecke, KB5TZY*
            Charles Battle
Text Design and Artwork by: *Plunk Design, Dallas, TX*
Cover Design by: *Sara Wright*

Acknowledgements
All photographs not credited are either courtesy of Author, Master Publishing,
Inc., Radio Shack or Howard W. Sams & Company.

Printed in the United States of America

9 8 7 6 5 4 3 2 1

Table of Contents

QUESTION POOL NOMENCLATURE

The latest nomenclature changes and question pool numbering system recommended by the volunteer examiner coordinator's question pool committee (QPC) for question pools effective July 1, 1993, have been incorporated in this book. The Novice Class (Element 2) and Technician Class (Element 3A) question pools have been rewritten at junior high school and high school reading levels, respectively. These question pools are valid from July 1, 1993 until June 30, 1997.*

* Per QPC: "Date may be superseded as a result of changes to the licensing structure, substantial rule changes, or the like."

Preface

Welcome to the fabulous hobby of Amateur Radio! Federal Communications Commission (FCC) action of February 14, 1991, made it easier than ever to obtain an entry-level license for the ham radio service. In just one test session, you may satisfy the requirements for obtaining the Technician Class no-code license. Absolutely no knowledge of Morse code is required!

The *Novice and Technician No-Code Plus* Amateur Radio license preparation book is designed specifically to help you prepare quickly for your entry-level license. If you decide to enter the amateur service without taking a code test, your no-code examination will consist of two simple written examinations. Both of these may be administered in the same session, one after the other. The first one is on Element 2 that applicants for the Novice Class license also take. The second is on Element 3A Technician Class theory. If you pass, you will be awarded a Technician Class no-code operator's license by the FCC.

Chapter 3 contains the 350-question Element 2 question pool, and the 295-question Element 3A pool. Element 2 questions, answers, and correct answer explanations are presented first, then Element 3A follows in the same format. *Everything* you need to prepare for these two written examinations for your Technician Class no-code license is provided in this book.

Another way to enter the amateur service is with a Novice Class license which requires you to pass an Element 2 written examination, as well as an Element 1A 5-wpm Morse code test. As mentioned, Chapter 3 contains the Element 2 question pool. Chapter 5 describes, in detail, simple steps to learn the Morse code, and suggestions for practicing Morse code.

Chapter 2 explains the privileges you will enjoy with a Technician Class no-code license. Chapters 4 and 6 give the privileges for the Novice and Technician-Plus Class licenses, respectively.

Chapter 7 explains how the examinations are conducted, the formats used, and what to expect at an examination session. An FCC application, Form 610, is included in the back of this book. Before you begin with the questions, answers, and descriptions for Novice, and then Technician, you will want to start at the very beginning of this book to determine which way you plan to enter the amateur service—either as a complete Novice or as a no-code Technician. The choice is yours.

Regardless of how you enter the ham radio ranks, thousands of amateur operators will be anxious to hear you on the air! By giving a small amount of effort, you should be ready for the tests in less than 30-40 days. Good luck!

Gordon West, WB6NOA

1

Amateur Service

ABOUT THIS BOOK

The purpose of the *Novice and Technican No-Code Plus* book is to prepare you so you may obtain an entry-level amateur operator/primary station license. You may enter as a licensed Novice operator by passing a Novice Element 2, 30-question written examination, and an Element 1A, simple 5-wpm code test. This leads you to many HF CW privileges, plus two bands on VHF/UHF.

Or you may enter the ranks of the amateur service without passing a code test with the Technician Class no-code license. You first need to pass a Novice Element 2, 30-question written examination, and then pass an Element 3A, 25-question written examination.. *But no code test is required!* This license allows you all amateur services privileges above 30 MHz.

After obtaining a Novice Class or Technician Class no-code license, you may want to upgrade to the Technician-Plus Class, then you will have both Novice and Technician Class privileges.

Since *Novice and Technician No-Code Plus* prepares applicants to pass the requirements of an amateur operator with Novice Class, or Technician Class no-code, and/or Technician-Plus privileges, the book chapters are organized accordingly. We have anticipated that readers may be interested in only certain chapters, and not the complete book. For this reason, the related chapters have been written to stand alone. This leads to some duplication for the person reading the complete book. We hope it is not excessive.

We will be talking about the different license classes in more detail, but before we get down to the different class privileges, let's spend a few minutes bringing you up to date on the exciting background of the amateur service, and the requirements of an amateur service license.

WHAT IS THE AMATEUR SERVICE?

More than 600,000 Americans are licensed amateur operators. According to our Federal Communications Commission (FCC), the U.S. Government agency responsible for licensing amateur operators, "The amateur service is for qualified persons of all ages who are interested in radio technique solely with a personal aim and without pecuniary interest." Ham radio, as it is known, is first, and foremost, *a fun hobby!* In addition, it is a service.

The amateur service exists under international treaty and is authorized in practically every country. Each government has its own rules for admission to the ham ranks. Numerous frequency bands throughout the radio spectrum are allocated to the amateur service on an international basis, making it possible for amateur operators to communicate with each other in all parts of the world—even in space. Astronaut Owen Garriott, W5LFL, became the first ham operator to operate from space during the Columbia/Spacelab mission in 1983 using a hand-held radio and an antenna in the window of the space shuttle. Gordon West, your author, was among the first hams to communicate with W5LFL in the Spacelab.

More than 2,000,000 operators exchange ham radio greetings and messages by voice, teleprinting, telegraphy, facsimile and television worldwide. Japan, which also has a no-code license, has over one million hams alone! It is very commonplace for U.S. amateurs to communicate with Russian amateurs, while mainland China is just getting started with their amateur service. Being a ham operator is a very good way to promote international good will.

AMATEUR SERVICE BENEFITS

The benefits of ham radio are countless! There is something for everyone. The ham operators are probably best known for their contributions during times of disaster. When all else fails, the ham operator traditionally gets through. In addition, over the years, amateurs have contributed much to electronic technology. They have even designed and built their own orbiting communications satellites.

On the other hand, amateur service isn't just for the technically inclined. It is for everyone of all ages. There are ham radio operators under eight years old—and many over eighty! Most hams are just plain folks, but many famous celebrities are amateur operators. Barry Goldwater is K7UGA, Jordan's King Hussein is JY1, ex-pro baseball player Joe Rudi is NK7U, guitarist Chet Atkins is WA4CZD, and singer Ronnie Milsap is WB4KCG. Other famous amateurs are Bill Halligan (who founded Hallicrafters), Arthur Godfrey, Andy Devine, Pee Wee Hunt, Alvino Rey, General Curtis LeMay . . . and recent Gordon West Radio School graduate, Priscilla Presley.

Ham radio for the handicapped is a godsend! It is a great equalizer, since everyone is the same behind a microphone or a packet radio computer keyboard. Being an amateur service ham operator can take the disabled to the far flung corners of the world. Recent FCC rules for the disabled may now allow special examination techniques to be offered anyone with a handicap. New FCC rules also allow for code test exemptions for General Class and Extra Class operators with a doctor-certified handicap.

The ham fraternity knows no geographic, political or social barrier. And if you stick to your guns, and spend the effort, you are going to be part of it. Probably the primary prerequisite for passing any amateur license examination is the will to do it. If you follow the suggestions in this book, your chances are excellent. If you are fascinated by radio communication, learning will be easy and fun!

LICENSE PRIVILEGES

An amateur operator license conveys many privileges. As the *control operator* of an amateur station, an amateur operator is responsible for the quality of the station's transmissions. Most radio equipment must be authorized by the government before it can be widely used by the public, but for the most part, not amateur equipment! Unlike the citizen's band service, amateurs may design, construct, modify and repair their own equipment. But you must have a license, and even though it is easier than ever, there are certain things you must know before you can obtain the needed license from the FCC. They are all covered in this book.

OPERATOR LICENSE REQUIREMENTS

To qualify for an amateur operator/primary station license, a person must pass an examination according to FCC guidelines. The degree of skill and knowledge the candidate can demonstrate to the examiners determines the class of operator license for which the person is qualified.

The new rules and regulations no longer require a Morse code test for the entry-level Technician Class license. While Morse code examinations are still in place for Novice Class (5-wpm), General Class (13 wpm), and Extra Class (20 wpm) operator licenses, you don't need to learn the code to get onto those exciting VHF and UHF line-of-sight frequencies plus the 6-meter band for worldwide communications. Once you're on the air as a Technician Class, you can think about learning the code at anytime.

There is a big difference between CB and amateur operation. The amateur service is both a public service and a hobby. CB exists for short-distance, low-power, personal and business communications only. There is no 150-mile distance limitation in ham radio as there is using CB. Using higher output power levels, you can talk around the world—limited only by radio propagation conditions. Hams are licensed with call signs and don't use "handles." Licensing and call signs were discontinued for the citizens band some years ago. CBers are prohibited from experimenting, which is the cornerstone and one of the fun activities of Amateur Radio. You'll find that ham radio offers you far more capabilities—more frequencies, higher power, more emissions, more modes—and lots of fun, too. Hams can even interconnect their radios with the telephone system.

Anyone is eligible to become a U.S. licensed amateur operator (including aliens, if they are not a representative of a foreign government). There is no age limitation. If you can pass the examinations, you can become a ham!

One of the reasons for the existence of amateur services is to provide communications in times of emergency. Although hardly ever used on the ham bands during an emergency anymore, CW (Morse code) is one way to pierce through interference when other modes cannot get through. Packet radio is another very special type of emergency communications, using your own home computer over the airwaves. Now that there is an opportunity to become an amateur operator Technician Class licensee, you can concentrate on packet communications without ever having to worry about learning the Morse code for emergency communications.

If you already know the Morse code, or will enjoy learning the code, then why not make your first license a complete Novice Class license — "complete" meaning you take both the Element 2 theory examination and the simple Element 1A 5-wpm code test, and pass both. That satisfies the complete requirements for Novice Class, and your examiners will send in the form to obtain your entry-level Novice Class license.

If you're not interested in learning the code, don't worry about it. There is no pressure to learn the code; you can take the code test at a later date. When you pass the code test, you will instantly become a Technician-Plus operator — meaning Technician Class "plus" the code.

OPERATOR LICENSE CLASSES

There are five official successive levels of amateur operator licenses. Each requires progressively higher levels of learning and proficiency, and each gives you additional operating privileges. There is no waiting time to upgrade from one amateur class to another, nor any required waiting time to retake a failed examination. You can take all of the examinations at one sitting if you want to.

This is what is known as *incentive licensing*—a method of strengthening the service by offering more privileges in exchange for more electronic knowledge and code skill. The theory and regulations covered in each of the questions of the various examinations relates to privileges that you will obtain when you upgrade. *Table 1-1* illustrates six levels of operating privileges for five classes of amateur operators. Note that the sixth level is the Technician-Plus Class, which is actually an extension of the Technician Class license. It means you have met the Technician Class requirements "plus" passed a code test.

Table 1-2 details the subjects covered in the various written examination question elements.

The distribution of the questions in each license class over the topics is shown in *Table 1-3*.

Table 1-1. Amateur License Classes and Exam Requirements

License Class	Test Element	Type of Examination
Novice Class	Element 2 Element 1A	30-Question Written Examination 5-wpm Code Test
[1,2] Technician Class	Element 2 and 3A	55-Question Written Examination (In 2 parts-30 Element 2, 25 Element 3A) (No Morse code requirement)
[2] Technician-Plus Class	Element 3A Element 1A	25-Question Written Examination if a Novice. 5-wpm Code Test if a Technician.
General Class	Element 3B Element 1B	25-Question Written Examination 13-wpm Code Test
Advanced Class	Element 4A	50-Question Written Examination (No additional Morse code requirement)
Extra Class	Element 4B Element 1C	40-Question Written Examination 20-wpm Code Test

[1] No-Code License
[2] Effective 2/14/91

Note: It is desirable that written examinations be taken in ascending order of difficulty all the way to Extra Class. Either Element 2 or Element 3A may be taken first since both are entry-level examinations. The code tests may be taken in *any* order. You can take the 20-wpm code test first if you can pass it. You can enter as a Technician without code, and then gain Technician-Plus CW privileges by passing the Element 1A code test. You can enter as a Novice with code, and then gain Technician-Plus by passing the Element 3A theory examination.

Table 1-2. Question Element Subjects

Element 2
    Novice
    Technician
	Elementary operating procedures, electronic theory and radio regulations

Element 3A
    Technician
    Technician-Plus
	Beginner-level amateur practices, electronic theory and radio regulations with VHF/UHF emphasis

Element 3B General	General operating privileges, amateur practices and radio regulations with emphasis on HF operation
Element 4A Advanced	Intermediate operating procedures, electronic theory and radio regulations
Element 4B Extra	Specialized operating procedures, electronic theory and radio regulations

Note: All license written examinations are additive. For example, to obtain a General Class license, you must take and pass an Element 2 written examination, an Element 3A written examination and an Element 3B written examination. It is better not to skip over a lower class written examination.

Table 1-3. Topic Distribution Over License Classes

Subelement	2 Novice Technician	3A Technician Tech-Plus	3B General	4A Advanced	4B Extra
1 Commission's Rules	10	5	4	6	8
2 Operating Procedures	2	3	3	1	4
3 Radio Wave Propagation	1	3	3	2	2
4 Amateur Radio Practices	4	4	5	4	4
5 Electrical Principles	4	2	2	10	6
6 Circuit Components	2	2	1	6	4
7 Practical Circuits	2	1	1	10	4
8 Signals and Emissions	2	2	2	6	4
9 Antennas and Feed Lines	3	3	4	5	4
Total questions	30	25	25	50	40

Two Entry-Level Classes

There are two beginner entry-level licenses—the traditional *Novice Class operator*, which requires a Novice theory examination and a code test, and the *Technician Class No-Code operator*, which requires two written examinations, but no code test. If you have been practicing the Morse code, or if you know the sounds of Morse code from the Boy Scouts or the military, then indeed plan to get started in ham radio by taking a code test along with the Novice (Element 2) theory material covered in this book.

However, if you are not interested in learning the Morse code, or have tried to learn the Morse code and just couldn't quite get an ear for it, then join us in the amateur service by going the way of a Technician Class no-code license.

Novice Class for Those That Enjoy Code

If you enjoy the Morse code, join the amateur service by preparing for your Novice Class license. The Novice Class elementary rules and theory examination, Element 2, and its 5-wpm beginner telegraphy test, Element 1A, make it the ideal license to receive amateur service call letters. The Novice Class license testing is now being folded into the volunteer examiner coordinator (VEC) System. Three accredited General or higher Class hams will form the team under the VEC System to give you the code and written examination. Since there are thousands of General Class hams (or higher) in all areas of the country, it won't be hard to find a team to give you your Novice Class entry-level examination.

To find a Novice Class VE team near you, refer to the Appendix for a list of volunteer examiner coordinators. Call a coordinator near you—or a national coordinator—and simply give them your zip code. They will help you make contact with a testing team that will be happy to give you an amateur operator examination at a location near you. Also, your Amateur Radio dealer and local Radio Shack stores likely will have information on examination sessions.

Technician Class No-Code

Another way to get into the amateur service is via a Technician Class no-code license. An applicant for this license must pass a written examination from Element 2 and a written examination from Element 3A. There is no code test. Both elements are contained in this book, and are easily passed with just a few weeks of going over the questions, the possible right answers, the correct answer, and the fun explanations.

The Technician written examinations must be taken in front of a team of three accredited volunteer examiners. In fact, now that the Novice Class testing program is being folded into the VEC System, you likely will be taking your Technician Class written examinations in front of the same volunteer examiners who would have administered your Novice Class examination if you had decided to go that entry-level route.

The Technician Class no-code license has been years in the making. The Federal Communications Commission opened up this new way of obtaining a no-code amateur service's license on February 14, 1991, after studying numerous petitions and public comments for an easier way to enter Amateur Radio. By offering a codeless class of license with privileges exclusively above 50 MHz, an entry-level license is finally available to those who find the Morse code a barrier to becoming a licensed amateur operator.

But the FCC decided to retain the current Novice Class operator license as an alternate entry-level license for those persons able to pass the 5-wpm Morse code test, instead of the more comprehensive written examination on Element 3A required for the Technician Class license.

"The amateur service is not growing as it should related to what it has to offer," commented FCC Private Radio Bureau Chief, Ralph Haller at a press conference right after the no-code license was announced. "...The amateur service is where our nation's technical expertise comes from—these changes should attract people who are interested in computers and digital communications, and should help the U.S. to become more competitive," adds Haller.

By granting the Technician Class operator all amateur privileges from 50 MHz on up, the new Technician Class operator will mainstream easily into the amateur service with little or no negative impact on present VHF and UHF operators. In fact, new Technician Class operators will help fill those bands that are vacant, and save us from possibly losing valuable frequencies in the future. "Use it or lose it" is the motto of amateur operators when looking at the vast expanse of Technician Class frequencies under-utilized throughout the country.

The response to a Technician Class license has been overwhelming. Gordon West's Radio School weekend classes have tripled in size, and these classes are now offered almost every weekend somewhere in the country.

Technician Class operators receive group "C" call signs where available, and this further helps them blend in with older operators who originally took a code test to get into the amateur service.

LICENSE UPGRADES

After entry into the amateur service either as a Novice Class or as a Technician Class, an amateur can progress up the ladder by passing the requirements of the more advanced license classes.

You may upgrade even before you receive a prior license! Your *Certificate of Successful Completion of Examination* (CSCE) from your examination session is your evidence that you have passed the necessary requirements. A photocopy of the CSCE showing that you have qualified for a new operator license must be given to the examiners when you apply for your upgrade. They will attach it to your Form 610 application as proof to the VEC that you are still awaiting your operator license.

The VET will also enter the date of issuance in the *Administering VE's Report* of your upgrade application. The VEC office will hold this application in a file because the VEC *cannot* forward your upgrade application to the FCC until you furnish them with a copy of your most previous operator license. The VECs are required to attach your most previous license to all new applications.

If you already have a call sign, you may immediately begin using your new privileges. For example, Novices who have received a CSCE for an upgrade to Technician (actually Technician-Plus) do not have to wait until the Technician license arrives to start operating on 2-meters! They can use their new privileges immediately. They must, however, append their call sign with an identifier to signify to everyone (especially the FCC who might be monitoring) that they have passed the necessary upgrade requirements, but have not yet received the license.

Novices who are waiting for their Technician license must include the letters "KT" preceded by either the slant bar (/) when transmitting in the telegraphy mode or the word "temporary" when making a voice transmission. No special identifier is required when operating within the privileges of your previous license. *Figure 1-1* shows is a list of the special temporary operating indicators.

Upgrade to	Temporary Identifier
Technician Class	KT*
General Calss	AG
Advanced Class	AA
Extra Class	AE

* A Technician Class no-code operator who upgrades to Technician-Plus need not use an identifier on newly obtained privileges since a new Technician license will not be issued. The CSCE validates the Technician-Plus operating privileges.

Figure 1-1. Temporary Identifiers for Upgrades

Technician-Plus Class

The "plus" added to the Technician Class license means "plus code." For Technician Class no-code operators to earn their Technician-Plus privileges, the optional 5-wpm Morse code test must be taken in front of a team of three accredited examiners. Note that we said *privileges* and not *license!* This is because no-code Technicians are not issued another Technician-Plus "license." Instead the CSCE issued by your VET to show you passed the necessary requirements authorizes your Technician-Plus privileges. It is the only Technician-Plus "license" that Technician Class no-code operators receive!

Novice Class operators, already proficient in 5-wpm Morse code, earn their Technician-Plus privileges by passing a VE-administered Technician Class Element 3A written examination.

Novices who pass Element 3A will be issued a Technician license by the FCC, but they still should keep a copy of their Novice license showing they have passed the 5-wpm code test. The Technician and Technician Plus "license" issued by the FCC looks (and is) exactly the same! It is up to the examinees to prove they have passed the necessary Element 1A code test with either a CSCE or a Novice license.

You may wish to re-read the comments about the Novice Class license; and if you know some code now, reconsider how you plan to enter ham radio. Maybe the best way is via the traditional Novice Class route by taking the Element 2 written theory examination and a code test in front of an accredited VE team of three General or higher Class ham buddies. After accomplishing that step, then at the same or another VE session, you can take the Element 3A written examination. Upon passing that examination, you are a Technician-Plus operator.

The big advantages of the Technician-Plus endorsement are Novice Class code and voice operating privileges *below* 30 MHz. This includes the popular voice portion on the 10-meter band, and the CW Novice privileges on 10 meters, 15 meters, 40 meters, and 80 meters. You may have heard about all the excitement on the 10-meter band for long-range voice communications, that should be incentive to you to up-grade to obtain Technician-Plus privileges.

General Class

The third step up the ladder is *General Class operator*. Another written examination (Element 3B) and demonstrated 13-wpm telegraphy skill (Element 1B) is required. This license authorizes all emission privileges on at least some portions of all amateur service frequency bands, including the very popular DX-oriented 20-meter ham band.

Advanced Class

The fourth step up the ladder is *Advanced Class operator*. More theory is required (Element 4A). This license authorizes additional frequency privileges on amateur service HF bands. There is no additional code requirement at this step.

Extra Class

About ten percent make it all the way to the top-of-the-line level, the *Amateur Extra Class operator*. Another written examination (Element 4B) and a 20-wpm telegraphy test (Element 1C) must be passed. This license authorizes all frequency privileges in all communications modes and emissions permitted to the amateur service. It offers you everything there is (or will be as technology develops) and is the ultimate goal of most ham operators.

AMATEUR CALL SIGNS

As an aid to enforcement of the radio rules, transmitting stations throughout the world are required to identify themselves at intervals when they are in operation. By international agreement, the prefix letters of a station's call sign indicates the country in which the station is authorized to operate. On the DX airwaves, hams can readily identify the national origin of the ham signal they hear by its call sign prefix. The national prefixes allocated to the United States are AA through AL, KA through KZ, NA through NZ, and WA through WZ. In addition, U.S. amateur stations with call signs that start with A, K, N or W are followed by a number indicating U.S. geographic area. *Table 1-4* details these geographical areas.

The suffix letters indicate a specific amateur station. The primary Amateur Radio call sign is issued by the FCC's licensing facility in Gettysburg, Pennsylvania, on a systematic basis after they receive your application from the VE team. A call sign is a very important matter to a ham—sometimes more personal than his name! Novice and some Technician call signs contain two prefix letters, a single radio district number (1 through 0) and three suffix letters. KA7ABC is an example.

Table 1-4. Call Sign Numbers for U.S.A. Geographical Areas

Call Sign Area No.	Geographical Area
1	Maine, New Hampshire, Vermont, Massachusetts, Rhode Island, Connecticut.
2	New York, New Jersey, and the U. S. Virgin Islands.
3	Pennsylvania, Delaware, Maryland, District of Columbia.
4	Virginia, North and South Carolina, Georgia, Florida, Alabama, Tennessee, Kentucky, Puerto Rico.
5	Mississippi, Louisiana, Arkansas, Oklahoma, Texas, New Mexico.
6	California, Hawaii.
7	Oregon, Washington, Idaho, Montana, Wyoming, Arizona, Nevada, Utah, Alaska.
8	Michigan, Ohio, West Virginia.
9	Wisconsin, Illinois, Indiana.
0	Colorado, Nebraska, North and South Dakota, Kansas, Minnesota, Iowa, Missouri.

EXAMINATION ADMINISTRATION

Until recently, any two hams who were General Class or higher, and 18 years old, could give an examination for a Novice Class license. That is now changing and all examinations are moving under the VEC System. The examinations are administered by three local amateur operators certified as volunteer examiners (VEs). VE teams usually provide the information on local VHF networks as to when and where examination sessions will be held. Their efforts are coordinated by a VEC who accredits them to serve as a volunteer examiner.

The questions for all examinations are developed and revised by the combined efforts of an internal committee consisting of VECs, called the Question Pool Committee, or QPC for short. The FCC no longer handles this function. Coordinating VECs make all questions and recommended multiple-choice answers available to the public. They are widely published by various license preparation material publishers. All VECs use the exact same worded questions and there are no secret questions for any class of amateur operator license.

Administering VEs may charge the candidate a test fee for certain reimbursable expenses incurred in preparing, processing and administering the amateur operator examination. The FCC annually adjusts the maximum amount of this test fee for inflation. After several adjustments, the fee is still under $6.00.

LICENSE EXAMINATIONS

The Technician Class no-code license examination consists of a two-part written examination. You will take both parts of the written examination in front of three volunteer examiners, usually at the same session. As mentioned previously, these volunteer examiners are accredited and registered through VECs to administer official amateur operator examinations. Presently, accredited volunteer examiners hold an Advanced Class or an Extra Class license, but soon General Class VEs will be accredited. They are not compensated; they perform the testing service on a voluntary basis. The VEs will help you fill out your FCC Form 610 (one is in the back of this book) and will help you properly process your paperwork with the Federal Communications Commission.

The Novice Class examination consists of an Element 2 written theory examination and Element 1A 5-wpm code test administered under the present system by two General or higher Class hams, 18 years or older. When the Novice testing is moved under the VEC system, the examination will be administered by an accredited team of three VEs. Passing the examination leads to a Novice Class license, and call letters will be issued to you by the Federal Communications Commission. This license permits CW privileges on certain HF frequencies, and voice privileges on certain VHF/UHF bands.

Yet another way to go is to obtain an entry-level license and upgrade privileges to Technician-Plus at the same time. To do this, you take and pass the Element 1A 5-wpm code test, Novice Element 2 written examination, and Technician Element 3A written examination—all in one session before a VE team. That's a lot of testing in just one session, but it's one more way that you might obtain your first amateur radio license.

STUDY TIME

If you are preparing for the Technician Class no-code license, it will take approximately 20 days to study Novice Element 2, and an additional 20 days to study Technician Element 3A. Remember, you should not skip over Novice Element 2 as you ascend the amateur service license ladder. All theory examination elements are designed to be additive.

If you already know the code, it will take you approximately 20 days to study for Novice Element 2, and a few days to brush up on your code at 5 wpm. If you don't know the code, Morse code practice tapes by your author are available from all Radio Shack stores and ham radio dealers. It takes about 30 days to memorize the Morse code practice by listening, and bring your speed up to the required 5 wpm.

Even though you might just take the two written examinations for the Technician Class no-code license, your examiners will also offer an optional 5-wpm code test. Why not give the test a try? You would end up with Technician-Plus Class privileges if you also pass the 5-wpm test!

The Question Pool

The Element 2 question pool contains 350 questions and the Element 3A question pool contains 295 questions. These questions, plus the multiple-choice answers, one of which is a right answer, will be the precise, exact questions used on your upcoming examination. The volunteer examiners will not re-word the questions, nor will they change any of the wrong or right answers. This is good news! No secret questions, no strange answers, and absolutely no surprises once you have covered this book. Letter for letter, word for word, and number for number—the precise questions and answers contained in this book are those to be found on your upcoming examinations.

Both the Element 2 and Element 3A question pools have been recently updated by the Question Pool Committee (QPC), under the leadership of Ray Adams, N4BAQ. The Question Pool Committee, and Adams, deserve tremendous credit in constantly keeping the question pool updated to agree with the latest FCC rules, and to match the current acceleration in amateur service technology with what is required to know for the examinations.

Chapter 3 is where you will find the two question pools. The Element 2 question pool is presented first, followed by the Element 3A question pool. When you take your Technician Class examination, all written, the Element 2 exam will usually be administered first, followed by the Element 3A exam. If you know these two question pools, you will pass the examinations with flying colors!

THE ACTUAL EXAMINATION

If you are preparing for the Technician Class no-code operator license, the examination will consist of 30 Novice Element 2 questions, and 25 Technician Element 3A questions, for a total of 55 questions divided between topic categories as shown in *Table 1-5*. The Element 2 examination will consist of 30 questions taken from the Element 2 350-question pool, each with its multiple-choice answers. The Element 3A examination will consist of 25 questions taken from the Element 3A 295-question pool, each with its multiple-choice answers. You must get at least 74 percent of the questions correct—22 of the 30 Element 2 questions and 19 of the 25 Element 3A questions—to pass.

If you plan to take the Novice CW test, read Chapter 5 for some very important hints.

TWO WAYS IN — YOUR CHOICE!

If you already know the Morse code, then you probably will want to enter the Amateur Radio hobby with a Novice license. If you haven't learned the code, and you are not interested in memorizing dots and dashes, then set your sights for the Technician Class no-code license.

Table 1-5. Question Distribution for the Technician Class No-Code Exam

Subelement	Number of Questions Element 2		Element 3A	
Commission's Rules (FCC rules for the Amateur Radio services)	N1	10	T1	5
Operating Procedures (Amateur station operating procedures)	N2	2	T2	3
Radio Wave Propagation (Radio wave propagation characteristics of amateur service frequency bands)	N3	1	T3	3
Amateur Radio Practices (Amateur Radio practices)	N4	4	T4	4
Electrical Principles (Electrical principles as applied to amateur station equipment)	N5	4	T5	2
Circuit Components (Amateur station equipment circuit components)	N6	2	T6	2
Practical Circuits (Practical circuits employed in amateur station equipment)	N7	2	T7	1
Signals and Emissions (Signals and emissions transmitted by amateur stations)	N8	2	T8	2
Antennas and Feed Lines (Amateur station antennas and feed lines)	N9	3	T9	3
TOTAL		30		25

Titles in parentheses are the official subelement titles listed in FCC Part 97.

You can even take your tests in steps! Once you pass an Element 2 examination, you have 365 days to complete either the Element 1A code test or the Element 3A examination. This means you don't have to take all of the examinations at once! You will receive certificates for successfully completing a portion of the requirements. The certificates are good for 365 days, so you have that much time to complete the remaining requirements for your license.

When taking a Technician Class no-code examination, if you fail Element 2, most VE teams will let you take the Element 3A portion. If you pass, you will receive a certificate showing you passed Element 3A. From that date you have 365 days to complete the Element 2 examination.

IT'S EASY!

Probably the primary prerequisite for passing any amateur operator license examination is the will to do it. If you follow the suggestions in this book, your chances are excellent. If you are fascinated by radio communication, and don't want to learn the code, now you have the opportunity to join the amateur service as a Technician Class no-code operator.

2

Technician Class No-Code Privileges

ABOUT THIS CHAPTER

There is plenty of excitement out there on the amateur service bands with your Technician Class license, either no-code or as a Technician-Plus Class up-grade. Just wait until you read about all of the band privileges you are going to get! No-Code privileges are covered in this chapter; Technician-Plus privileges are covered in Chapter 6.

TECHNICIAN CLASS NO-CODE LICENSE

The Technician Class no-code license is now the most popular way to get started in the amateur service. By taking two simple written examinations and scoring at least 74% on each exam, you will pass and be awarded a Technician Class operator's license. You will not need to take any type of code test to obtain your Technician Class license.

The Technician Class no-code license privileges begin at 50 MHz in frequency and go higher from there. As shown in *Table 2-1*, you have full operating privileges on the 6-meter worldwide band; the 2-meter band, which is the world's most popular repeater band; the 222-MHz band where repeaters are linked together; the 440-MHz band where you can operate amateur television, satellite, and remote-base stations; and the 1270-MHz band with more amateur television, satellite communications, and repeater linking. On top of this, there are several other microwave bands on which you can experiment—all this, without a code test!

TECHNICIAN-PLUS CLASS UPGRADE

Remember, there are two paths to the Technician-Plus Class operator upgrade. If you are a Novice Class operator, all you need to do is pass an Element 3A 25-question written examination at a VE test session to have Technician-Plus Class privileges.

If you enter the amateur service as a Technician Class, all you need to do is pass a 5-wpm Morse code (Element 1A) test at a VE test session, and you will have Technician-Plus Class privileges.

The Technician-Plus Class operator has privileges on all frequency bands at and above 50 MHz, just like the Technician Class no-code operator. But the Technician-Plus Class operator also receives 10-meter worldwide voice and digital privileges, 10-meter CW privileges, 15-meter CW privileges, 40-meter CW privileges, and 80-meter CW

Table 2-1. Technician Class No-Code Operating Privileges

Wavelength Band	Frequency	Emissions	Comments
160 Meters	1800–2000 kHz	None	No privileges
80 Meters	3675–3725 kHz	CW	No privileges without code test
40 Meters	7100–7150 kHz	CW	No privileges without code test
30 Meters	10100–10150 kHz	None	No privileges
20 Meters	14000–14350 kHz	None	No privileges
17 Meters	18068–18168 kHz	None	No privileges
15 Meters	21100–21200 kHz	CW	No privileges without code test
12 Meters	24890–24990 kHz	None	No privileges
10 Meters	28100–28500 kHz	CW, Data	No privileges without code test
	28300–28500 kHz	Voice	No privileges without code test
6 Meters	50.0–54.0 MHz	All modes	Sideband voice, radio control, FM repeater, digital computer, remote bases, and autopatches. Even CW. (1500 watts PEP output)
2 Meters	144–148 MHz	All modes	All types of operation including satellite and owning repeater and remote bases. (1500 watt PEP output)
1¼ Meters	222–225 MHz	All modes	All band privileges. (1500 watt PEP output)
70 cm	420–450 MHz	All modes	All band privileges, including amateur television, packet, RTTY, FAX, and FM voice repeaters. (1500 watt PEP output.)
33 cm	902–928 MHz	All modes	All band privileges. Plenty of room! (1500 watt PEP output.)
23 cm	1240–1300 MHz	All modes	All band privileges. (1500 watt PEP output)

privileges. 10 meters, 15 meters, 40 meters, and 80 meters are those worldwide bands that let you communicate anywhere around the world, anytime, day or night, with a worldwide set.

TECHNICIAN CLASS NO-CODE OPERATING PRIVILEGES

But don't be disappointed if you put off the code test—the Technician Class no-code operator has plenty of worldwide excitement on 6 meters, the first band we'll explore in detail.

6-METER WAVELENGTH BAND, 50.0-54.0 MHz

The Technician Class no-code operator will enjoy all amateur service privileges and maximum output power of 1500 watts on this worldwide band. Are you into radio control (R/C), and want to escape the interference between 72 and 76 MHz? On 6 meters, your Technician Class license allows you to operate on exclusive radio control channels at 50 MHz and 53 MHz, just for licensed hams.

On 6 meters, the Technician Class no-code operator can get a real taste of long-range skywave skip communications. During the summer months, and during selected days and weeks out of the year, 50-54

MHz, 6-meter signals are refracted by the ionosphere, giving you incredible long-range communication excitement. It's almost a daily phenomena during the summer months for 6 meters to skip all over the country. This is the big band for the Technician Class no-code operator because of this type of ionospheric, long-range, skip excitement. There are even repeaters on 6 meters. So make 6 meters "a must" at your future operating station. *Table 2-2* shows the ARRL 6-meter wavelength band plan:

Table 2-2. ARRL* 6-Meter Wavelength Band Plan, 50.0-54.0 MHz

MHz	Use
50.100–50.300	SSB, CW
50.100–50.125	DX window
50.110	SSB calling frequency
50.300–50.600	Non-voice communications
50.620	Digital/Packet calling frequency
50.800–50.980	Radio control
	20 kHz channels
51.000–51.100	Pacific DX window
51.120–51.480	Repeater inputs (19)
51.120–51.180	Digital repeater inputs
51.620–51.980	Repeater outputs (19)
51.620–51.680	Digital repeater outputs
52.000–52.480	Repeater inputs (23)
52.020, 52.040	FM simplex
52.500–52.980	Repeater outputs (23)
52.525, 52.540	FM simplex
53.000–54.480	Repeater inputs (19)
53.000, 53.020	FM simplex
53.1/53.2/53.3/53.4**	Radio control**
53.500–53.980	Repeater outputs (19)
53.5/53.6/53.7/53.8**	Radio control**
53.520	Simplex
53.900	Simplex

*American Radio Relay League   **Optional, local choice

2-METER WAVELENGTH BAND, 144 MHz-148 MHz

The 2-meter band is the world's most popular spot for staying in touch through repeaters. Here is where most all of those hand-held transceivers operate, and the Technician Class no-code operator receives unlimited 2-meter privileges!

Handie-talkie channels	Simplex autopatch	Rag-chewing
Transmitter hunts	Contests	Radio teleprinter
Autopatch	Traffic handling	Radio facsimile
Moon bounce	Satellite downlink	Emergency nets
Meteor bursts	Satellite uplink	Sporadic-E DX
Packet radio	Remote base	Aurora
Tropo-DX-ducting	Simplex operation	

The United States, and many parts of the world, are blanketed with clear, 2-meter, repeater coverage. They say there is nowhere in the United States you couldn't reach at least one or two repeaters with a little hand-held transceiver. 2-meters has you covered!

The Technician Class no-code license allows 1500 watts maximum power output for specialized 2-meter communications, and also permits you to own and control a 2-meter repeater. *Table 2-3* gives the 2-meter wavelength band plan proposed by the ARRL VHF/UHF advisory committee:

Table 2-3. ARRL 2-Meter Wavelength Band Plan, 144-148 MHz

MHz	Use
144.00–144.05	EME (CW)
144.05–144.06	Propagation beacons (old band plan)
144.06–144.10	General CW and weak signals
144.10–144.20	EME and weak-signal SSB
144.20	National SSB calling frequency
144.20–144.30	General SSB operation, upper sideband
144.275–144.300	New beacon band
144.30–144.50	New OSCAR subband plus simplex
144.50–144.60	Linear translator inputs
144.60–144.90	FM repeater inputs
144.90–145.10	Weak signal and FM simplex
145.10–145.20	Linear translator outputs plus packet
145.20–145.50	FM repeater outputs
145.50–145.80	Miscellaneous and experimental modes
145.80–146.00	OSCAR subband—satellite use only!
146.01–147.37	Repeater inputs
146.40–146.58	Simplex
146.61–146.97	Repeater outputs
147.00–147.39	Repeater outputs
147.42–147.57	Simplex
147.60–147.99	Repeater inputs

Repeater frequency pairs (input/output):

144.61/145.21	144.89/145.49	146.40 or 146.60/147.00*
144.63/145.23	146.01/145.61	146.43 or 146.63/147.03*
144.65/145.25	146.04/146.64	146.46 or 146.66/147.06*
144.67/145.27	146.07/146.67	147.69/147.09
144.69/145.29	146.10/146.70	147.72/147.12
144.71/145.31	146.13/146.73	147.75/147.15
144.73/145.33	146.16/146.76	147.78/147.18
144.75/145.35	146.19/146.79	147.81/147.21
144.77/145.37	146.22/146.82	147.84/147.24
144.79/145.39	146.25/146.85	147.87/147.27
144.81/145.41	146.28/146.88	147.90/147.30
144.83/145.43	146.31/146.91	147.93/147.33
144.85/145.45	146.34/146.94	147.96/147.36
144.87/145.47	146.37/146.97	147.99/147.39

Some states use a different band plan. Additional channels available in large cities when 15 kHz and 20 kHz "splinter channels," interspersed between regular channels, are used.
*local option

1¼-METER WAVELENGTH BAND, 222 MHz-225 MHz

Table 2-4 shows the proposed usage for this band. It is filled with activity by Novice Class operators transmitting from 222.1 MHz to 223.91 MHz. The Technician Class no-code license permits you to use the entire band, and 1500 watts maximum output power instead of the 25 watts for Novice Class licensees. If you need some relief from the

Table 2-4. ARRL 1¼-Meter Wavelength Band Plan, 222-225 MHz

MHz	Use
222.00–222.15	Weak-signal modes
222.00–222.05	EME
222.05–220.06	Propagation beacons
222.10	SSB and CW calling frequency
222.10–222.15	Weak signal CW and SSB
222.15–222.25	Local coordinator's option:
	Weak signal, ACSB, repeater inputs, control points
222.25–223.38	FM repeater inputs only
223.40–223.52	FM simplex
223.5	Simplex calling frequency
223.52–223.64	Digtal, packet
223.64–223.7	Links, contol
223.71–223.85	Local coordinator's option:
	FM simplex, packet, repeater outputs
223.85–224.98	Repeater outputs only

Simplex frequencies (MHz):

223.42	223.52	223.62	223.72	223.82
223.44	223.54	223.64	223.74	223.84
223.46	223.56	223.66	223.76	223.86
223.48	223.58	223.68	223.78	223.88
223.50*	223.60	223.70	223.80	223.90

*National simplex frequency

Repeater frequency pairs (input/output)(MHz):

222.32/223.92	222.54/224.14	222.76/224.36	222.98/224.58	223.20/224.80
222.34/223.94	222.56/224.16	222.78/224.38	223.00/224.60	223.22/224.82
222.36/223.96	222.58/224.18	222.80/224.40	223.02/224.62	223.24/224.84
222.38/223.98	222.60/224.20	222.82/224.42	223.04/224.64	223.26/224.86
222.40/224.00	222.62/224.22	222.84/224.44	223.06/224.66	223.28/224.88
222.42/224.02	222.64/224.24	222.86/224.46	223.08/224.68	223.30/224.90
222.44/224.04	222.66/224.26	222.88/224.48	223.10/224.70	223.32/224.92
222.46/224.06	222.68/224.28	222.90/224.50	223.12/224.72	223.34/224.94
222.48/224.08	222.70/224.30	222.92/224.52	223.14/224.74	223.36/224.96
222.50/224.10	222.72/224.32	222.94/224.54	223.16/224.76	223.38/224.98
222.52/224.12	222.74/224.34	222.96/224.56	223.18/224.78	

activity on 2 meters, the 222-225 MHz band is similar in propagation and use.

The Federal Communications Commission has reallocated the bottom 2 MHz of this band, 220-222 MHz, to the land mobile commercial radio service. The 220-222-MHz band is a shared band and narrow-band business radio is not compatible with amateur transmissions. It is hoped that the Technician Class no-code license will fill all the VHF and UHF bands with activity so we don't ever lose more frequencies due to inactivity!

70-cm WAVELENGTH BAND, 420-450 MHz

As you gain more experience on the VHF and UHF bands, you will soon be invited to the upper echelon of specialty clubs and organizations. The 450-MHz band is where the experts hang out. Amateur television (ATV) is very popular, so there's no telling who you may *see* as well as hear. This band also has the frequencies for controlling repeater stations and base stations on other bands, plus satellite activity. With a Technician Class no-code license, you may even be able to operate on General Class worldwide frequencies if a General Class or higher control operator is on duty at the base control point. You would be able to talk on your 450-MHz hand-held transceiver and end up in the DX portion of the 20-meter band. As long as the control operator is on duty at the control point, your operation on General Class frequencies is completely legal!

The 450 MHz band is also full of packet communications, RTTY, FAX, and all those fascinating FM voice repeaters. If you are heavy into electronics, you'll hear fascinating topics discussed and digitized on the 450-MHz band. A Technician Class no-code operator has full power privileges as well as unrestricted emission privileges. *Table 2-5* presents the ARRL 70-cm (centimeter) wavelength band plan.

33-cm WAVELENGTH BAND, 902-928 MHz

Radio equipment manufacturers are just beginning to market equipment for this band. Many hams are already on the air using home-brew equipment for a variety of activities. If you are looking for a band with the ultimate in elbow room, this is it!

Table 2-6 shows the 33-cm wavelength band plan adopted by the ARRL Board of Directors in July, 1989.

Table 2-5. ARRL 70-cm Wavelength Band Plan, 420-450 MHz

MHz	Use
420.00–426.00	ATV repeater or simplex with 421.25-MHz video carrier control links and experimental
426.00–432.00	ATV simplex with 427.250-MHz video carrier frequency
432.00–432.07	EME (Earth-Moon-Earth)
432.07-432.08	Propagation beacons (old band plan)
432.08–432.10	Weak-signal CW
432.10	70-cm calling frequency
432.10–433.00	Mixed-mode and weak-signal work
432.30-432.40	New beacon band
433.00–435.00	Auxiliary/repeater links
435.00–438.00	Satellite only (internationally)
438.00–444.00	ATV repeater input with 439.250-MHz video carrier frequency and repeater links
442.00–445.00	Repeater inputs and outputs (local option)
445.00–447.00	Shared by auxiliary and control links, repeaters and simplex (local option); (446.0-MHz national simplex frequency)
447.00–450.00	Repeater inputs and outputs

Repeater frequency pairs (input/output is local option)(MHz):

442.000/447.000	442.600/447.600	443.200/448.200	443.800/448.800	444.400/449.400
442.025/447.025	442.625/447.625	443.225/448.225	443.825/448.825	444.425/449.425
442.050/447.050	442.650/447.650	443.250/448.250	443.850/448.850	444.450/449.450
442.075/447.075	442.675/447.675	443.275/448.275	443.875/448.875	444.475/449.475
442.100/447.100	442.700/447.700	443.300/448.300	443.900/448.900	444.500/449.500
442.125/447.125	442.725/447.725	443.325/448.325	443.925/448.925	444.525/449.525
442.150/447.150	442.750/447.750	443.350/448.350	443.950/448.950	444.550/449.550
442.175/447.175	442.775/447.775	443.375/448.375	443.975/448.975	444.575/449.575
442.200/447.200	442.800/447.800	443.400/448.400	444.000/449.000	444.600/449.600
442.225/447.225	442.825/447.825	443.425/448.425	444.025/449.025	444.625/449.625
442.250/447.250	442.850/447.850	443.450/448.450	444.050/449.050	444.650/449.650
442.275/447.275	442.875/447/875	443.475/448.475	444.075/449.075	444.675/449.675
442.300/447.300	442.900/447.900	443.500/448.500	444.100/449.100	444.700/449.700
442.325/447.325	442.925/447.925	443.525/448.525	444.125/449.125	444.725/449.725
442.350/447.350	442.950/447.950	443.550/448.550	444.150/449.150	444.750/449.750
442.375/447.375	442.975/447.975	443.575/448.575	444.175/449.175	444.775/449.775
442.400/447.400	443.000/448.000	443.600/448.600	444.200/449.200	444.800/449.800
442.425/447.425	443.025/448.025	443.625/448.625	444.225/449.225	444.825/449.825
442.450/447.450	443.050/448.050	443.650/448.650	444.250/449.250	444.850/449.850
442.475/447.475	443.075/448.075	443.675/448.675	444.275/449.275	444.875/449.875
442.500/447.500	443.100/448.100	443.700/448.700	444.300/449.300	444.900/449.900
442.525/447.525	443.125/448.125	443.725/448.725	444.325/449.325	444.925/449.925
442.550/447.550	443.150/448.150	443.750/448.750	444.350/449.350	444.950/449.950
442.575/447.575	443.175/448.175	443.775/448.775	444.375/449.375	444.975/449.975

Table 2-6. ARRL 33-cm Wavelength Band Plan, 902-928 MHz

MHz	Use
902.0–903.0	Weak signal (902.1 calling frequency)
903.0–906.0	Digital Communications (903.1 alternate calling frequency)
906.0–909.0	FM repeater inputs
909.0–915.0	ATV
915.0–918.0	Digital Communications
918.0–921.0	FM repeater outputs
921.0–927.0	ATV
927.0–928.0	FM simplex and links

23-cm WAVELENGTH BAND, 1240-1300 MHz

There is plenty of over-the-counter radio equipment for this band, thanks to Novice Class operators getting onto the frequencies and exploring how far microwave signals go. Although the Novice Class operator is allowed only 5 watts, the Technician Class no-code operator may run any amount of power—with 20 watts about the maximum limit.

The frequencies are in the microwave region, and this band is excellent to use with local repeaters in major cities.

Like the 450-MHz band and the 2-meter band, this band is sliced into many specialized operating areas. You can work orbiting satellites, operate amateur television, or own your own repeater with your Technician Class no-code license. *Table 2-7* presents the 23-cm wavelength band plan adopted by the ARRL Board of Directors in January, 1985.

10-GHz (10,000 MHz!) BANDS AND MORE

There are several manufacturers of ham microwave transceivers and converters for this range, so activity is excellent. Gunnplexers are the popular transmitter. Using horn and dish antennas, 10 GHz is frequently used by hams to establish voice communications for controlling repeaters over paths from 20 miles to 100 miles. Output power levels are usually less than one-eighth of a watt! It's really fascinating to see how directional the microwave signals are. If you live on a mountaintop, 10 GHz is for you.

All modes and licensees except Novices are authorized on the bands shown in *Table 2-8*. There is much Amateur Radio experimentation on these bands.

Table 2-7. ARRL 23-cm Wavelength Band Plan, 1240-1300 MHz

MHz	Use
1240–1246	ATV #1
1246–1248	Narrow-bandwidth FM point-to-point links and digital, duplex with 1258-1260 MHz
1248–1252	Digital communications
1252–1258	ATV #2
1258–1260	Narrow-bandwidth FM point-to-point links and digital, duplexed with 1246-1252 MHz
1260–1270	Satellite uplinks, reference WARC '79
1260–1270	Wide-bandwidth experimental, simplex ATV
1270–1276	Repeater inputs, FM and linear, paired with 1282-1288 MHz, 239 pairs every 25 kHz, e.g., 1270.025, 1270.050, 1270.075, etc. 1271.0-1283.0 MHz uncoordinated test pair
1276–1282	ATV #3
1282–1288	Repeater outputs, paired with 1270-1276 MHz
1288–1294	Wide-bandwidth experimental, simplex ATV
1294–1295	Narrow-bandwidth FM simplex services, 25-kHz channels
1294.5	National FM simplex calling frequency
1295–1297	Narrow bandwidth weak-signal communications (no FM)
1295.0–1295.8	SSTV, FAX, ACSB, experimental
1295.8–1296.0	Reserved for EME, CW expansion
1296.0–1296.05	EME exclusive
1296.07–1296.08	CW beacons
1296.1	CW, SSB calling frequency
1296.4–1296.6	Crossband linear translator input
1296.6–1296.8	Crossband linear translator output
1296.8–1297.0	Experimental beacons (exclusive)
1297–1300	Digital communications

Table 2-8 Gigahertz Bands

2.30–2.31 GHz	10.0–10.50 GHz*	165.0–170.0 GHz
2.39–2.45 GHz	24.0–24.25 GHz	240.0–250.0 GHz
3.30–3.50 GHz	48.0–50.00 GHz	All above 300 GHz
5.65–5.925 GHz	71.0–76.00 GHz	

*Pulse not permitted

SUMMARY

Enjoy the excitement of being a ham radio operator and communicate worldwide with other amateur operators without a code test. Your Technician Class no-code operator license allows you all ham operator privileges on all bands with frequencies greater than 50 MHz. You have operating privileges on the worldwide 6-meter band, on the world's popular 2-meter and 222-MHz repeater bands, on the amateur television, satellite communications, and repeater linking 440-MHz and 1270-MHz bands, and on the line-of-sight microwave bands at 10 GHz and above.

And if you know some Morse code, you can pass an Element 1A 5-wpm code test and become a Technician-Plus Class operator with voice privileges on 10 meters, and long-range CW privileges on 10 meters, 15 meters, 40 meters and 80 meters.

AN IMPORTANT WORD ABOUT
SHARED AND PROTECTED FREQUENCIES

Chances are that the privileges we have indicated in this chapter are the ones you will receive. It is important, however, that you know that every ham band above 225 MHz is shared on a secondary basis with other services. This means that the primary users get first claim to the frequency!

For example, Government radiolocation (radar) is a primary user on some bands. And a multitude of industrial, scientific and medical services have access to the 902-928-MHz band. Just because the frequency is allocated to the amateur service, does not mean that others do not have prior right or an equal right to the spectrum. *You must not interfere with other users of the band.*

There are also instances where amateurs must not cause interference to other stations, such as, foreign stations operating along the Mexican and Canadian border, military stations near military bases, and FCC monitoring stations. Also, amateur operators must not cause interference in the so-called *National Radio Quiet Zone* which is near radio astronomy locations. The astronomy locations are protected by law from Amateur Radio interference. Operation aboard ships and aircraft is also restricted. The FCC can curtail the hours of your operation if you cause general interference to the reception of telephone, or radio/TV broadcasting by your neighbors.

Every amateur should have a copy of the Amateur Radio Service Part 97 Rules and Regulations. It would be good for you to especially read Part 97.303 on frequency sharing requirements.

3

Getting Ready for the Examination

ABOUT THIS CHAPTER

Your examinations for the Technician Class no-code license will be taken from an Element 2 350-question pool, and an Element 3A 295-question pool. Each question in the pools has multiple-choice answers.

The written Element 2 examination will contain exactly 30 questions. You must get 74% of the answers correct. That means that you must answer 22 questions correctly, or miss no more than 8 questions.

On the Element 3A examination, 25 questions will be selected word-for-word from the 295-question pool. You must score at least 74%, which means you must answer 19 questions correctly, or miss no more than 6 questions.

For the Technician Class no-code license, the Element 2 and Element 3A examinations may be administered one after another, in ascending order. Although not required, you should pass Element 2 before you are administered Element 3A. If you fail the Element 2 examination, your testing team may allow you to try another examination immediately, or allow you to go on to Element 3A.

You may take both elements at once, or possibly on separate days, if it fits the schedule of the test team. Or you may take them separately within 365 days to complete your entry-level Technician no-code license.

For your Novice Class examination, the Element 2 written examination or the Element 1A 5-wpm code test may be taken in any order; however, the order will probably depend on how the VE team has the examination session scheduled. If you pass the written theory examination, but fail the code, you still receive credit for passing the Element 2 examination, and have 365 days to pass the code test. Similar conditions apply if you pass the code test and not the written examination. The requirements to pass the Novice written examination are the same as mentioned above, you must get at least 74 percent of the answers correct, that means you must answer 22 of the 30 questions correctly. The Element 1A code test will be similar to what is described in Chapter 5, which discusses what is in the test and what you need to do to pass it.

The Technician-Plus Class examination will depend on your entry path into the amateur service. If you are a Novice Class operator, you will be administered the 25-question Element 3A written examination.

If your are a Technician Class no-code operator, you will be administered an Element 1A 5-wpm code test. Passing the required written
examination or code test will result in your receiving Technician-Plus
Class privileges. Of course, if you decide to satisfy all the requirements
in one sitting, and you pass the Element 2 and Element 3A written
examinations, and the Element 1A code test, you will enjoy Technician-
Plus operator privileges.

All examinations are administered by volunteer examiners—
amateur radio operators that are accredited by an area or national
volunteer examination coordinator. You will receive a Certificate of
Successful Completion of Examination (CSCE) for passing any portion
of a multiple-part examination. As suggested previously, if you fail an
examination, whether it be code or theory, your testing team may
allow you to try another examination immediately, or allow you to go
on to the next higher element.

The questions and answers will be identical to those that follow in
this chapter. Each question will be word-for-word, with four possible
word-for-word, multiple-choice answers, one of which will be correct.
They will not vary from what is published in this book. The only thing
you won't find on the test is the specified correct answer and explanation that are given in this book!

WHAT THE EXAMINATIONS CONTAIN

The examination questions and the multiple-choice distractors for all
license class levels are public information. They are widely published,
and are identical to those in this book. FCC rules prohibit any examiner or examination team from making *any changes* to any questions,
including any numerical values. No numbers, words, letters, or punctuation marks can be altered from the published question pool. By
studying the Element 2 and Element 3A question pools in this book,
you will be reading the same exact questions that will appear on your
30-question Element 2 written examination, and your 25-question
Element 3A examination.

Table 3-1 shows how the Element 2 30-question examination and
Element 3A 25-question examination will be selected from the two
question pools. Each pool is divided into various subelements. Each
subelement covers a different subject. For example, for the Element 2
examination, 10 questions of the 30 total will be taken from the
Commission's Rules subelement. For the Element 3A examination, 5
questions of the 25 total will be taken from the Commission's Rules
subelement. There are a total of 113 questions in the Commission's
Rules subelement of Element 2's question pool from which the VETs
can select 10 questions for your examination. While in Element 3A,
there are 55 questions in the same subelement from which the VETs
can select five questions for your examination.

All volunteer examination teams use the same pool of questions
and multiple-choice answers that are in this chapter. This uniformity

in study material insures common examinations throughout the country. Most examinations are computer-generated, and the computer selects the right number of questions out of each subelement for your two upcoming examinations.

Table 3-1. FCC Question Pool

Element 2 Subelement		Page	Total Questions	Examination Questions
N1	Commission's Rules	33	113	10
N2	Operating Procedures	60	35	2
N3	Radio Wave Propagation	68	12	1
N4	Amateur Radio Practices	71	44	4
N5	Electrical Principles	82	44	4
N6	Circuit Components	92	23	2
N7	Practical Circuits	97	24	2
N8	Signals and Emissions	103	22	2
N9	Antennas and Feed Lines	109	33	3
	TOTALS		350	30
Element 3A Subelement		Page	Total Questions	Examination Questions
T1	Commission's Rules	118	55	5
T2	Operating Procedures	131	41	3
T3	Radio Wave Propagation	141	33	3
T4	Amateur Radio Practices	150	53	4
T5	Electrical Principles	163	22	2
T6	Circuit Components	169	25	2
T7	Practical Circuits	176	11	1
T8	Signals and Emissions	179	22	2
T9	Antennas and Feed Lines	184	33	3
	TOTALS		295	25

All amateur service examination questions are reviewed periodically by a team of volunteer examination coordinators assembled as a Question Pool Committee (QPC). These are fellow hams with Extra Class licenses who volunteer their time to insure that the current tests are accurate and fair, and are subject matter relevant. As technology changes, so will the questions. A public notice will be given approximately one year in advance of any question changes so publishers can revise their study materials.

Question Coding

Each question of the 350 Element 2 Novice Class question pool and the 295 Element 3A Technician Class question pool, is coded according to the numbers and letters assigned to each question. The coded numbers and letters reveal important facts about each question. The numbering code was completely revised on July 1, 1993. The code always contains five alphanumeric characters to identify the question. This numbering system will completely replace the old numbering system in all of the question pools as the QPC revises pools.

In *Figure 3-1*, a number from the Element 3A pool is used as an example. T5A02 means the following: The first character identifies the class of license for which the question pool is to be used. The second digit identifies the subelement number. As shown in *Table 3-1*, for the Novice Class (Element 2) questions, the first character is always "N"; for the Technician Class (Element 3A) questions, the first character is always "T". (For other classes of license, it is G for General, A for Advanced, and E for Extra.) The second digit (1-9) specifies the particular subelement that the question emphasizes ("5" in this example indicates "Electrical Principles"). The third character indicates the topic within the subelement (Topic "A" in this example). The fourth and fifth digits indicate the question number within the subelement topic's group of questions ("02" in this example specifies the second question). Note that the question number is always two digits with a leading zero as required.

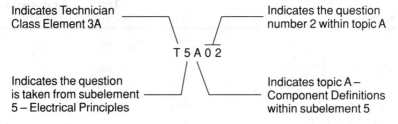

Figure 3-1. Examination Question Coding

NEW JULY 1, 1993 QUESTION POOLS

This book has been revised to include the new July 1, 1993 question pools. The Novice Class (Element 2) and Technician Class (Element 3A) question pools have been carefully designed to meet the reading and understanding requirements of a junior high school and high school level of comprehension, respectively. Members of the VEC Question Pool Committee (QPC), assisted by inputs from the amateur community and the compiling efforts of Steve Sternitzke, NS5I, have spent hundreds of hours developing these new questions. These changes will make it easier for young adults to enter the amateur service.

The QPC members are always interested in receiving input to their question pools. If you have suggestions for improving the Novice Class and Technician Class questions, valid July 1, 1993, please write your comments to your author (address in the Appendix) who will then pass them to the QPC for review.

A QUICK LOOK AT EXAM QUESTIONS

Go ahead now and scan through the Element 2 and Element 3A questions. As you scan through, note the subelements we talked about. Note also that the number of examination questions to be taken from the subelement is given. This corresponds to the number in *Table 3-1* for the subelement. Notice that the question coding, which we explained, indicates when subelements and topics change.

Be aware that there is only one correct answer. Notice also that the correct answer and the correct answer explanation are given right after the question so that you have an easy time learning the material.

HINTS ON STUDYING FOR THE EXAMS

First decide which license you are going to tackle—the Novice Class license requiring the Element 2 theory and Element 1A code, or the Technician Class license requiring the Element 2 theory and the Element 3A theory.

Code Study

If you are going to learn the code, or review your knowledge of the code, we recommend the following code study:
1. Use the Radio Shack (62-2418) learning code and code speed building cassettes recorded by your author. These are a fun way to learn Morse code.
2. Use Morse code learning software programs on your personal computer! These have the advantage of generating unique code text and random characters that cannot be easily memorized. Some have code learning and proficiency games that make telegraphy fun! You also can adjust these programs to transmit code at any speed, character spacing, and tone frequency—even take computer-constructed telegraphy examinations right at the keyboard!
3. Use Radio Shack and ham dealer computer code courses.
4. Attend classroom code practice instruction.
5. Listen and copy from high frequency radio with CW capabilities.
6. Practice code with a buddy at your house with a code-practice oscillator.

No-Code Study

Here are some suggestions to make your learning easier:
1. Begin with Subelement 1 of each of the Element 2 and Element 3A question pools. Study the questions, the correct answer, and the brief explanation. Be sure to look at the incorrect answers and see how close they are to the right one. Element 2 is presented first, then Element 3A.

Element 2 begins with Subelement N1—Commission's Rules. Element 3A begins with Subelement T1—Commission's Rules. The

Element 2 question pool contains 350 questions and the Element 3A question pool contains 295 questions.

If you are planning to obtain the Novice License with code, begin practicing your code learning skills.

2. Since the Technician Class requires two written examinations, it may be best to first study Element 2, and then have someone give you a trial examination according to the distribution shown in *Table 3-1*. Once you have Element 2 down cold, you can move on to Element 3A to complete your study for both written examinations. You usually will be required to take the Element 2 examination before you take the Element 3A examination.

   If you think you might forget what you learned in Element 2, contact an accredited VEC examination team and arrange to take Novice Element 2 all by itself. If you pass Element 2, you will receive a CSCE (Certificate of Successful Completion of Examination), and this will allow you 365 days to pass either the Element 3A written examination to become a Technician Class no-code operator, or to pass the Novice Class Element 1A code test to completely validate the Novice Class license.

   In the Gordon West Radio School classes, Element 2 is taught first, then an outside examination team comes in and tests the students just on Element 2. Next, the school concentrates on Element 3A.

3. Be sure to read over each multiple-choice answer carefully. Some start out looking good, but just one or two words may change the answer from right to wrong. Also, don't anticipate that the multiple-choice answers will always appear in the exact same A-B-C-D order on your actual Technician Class examinations. They may not.

4. Keep in mind how many questions may be taken out of any one subelement.

5. A fun way of preparing for the exam is to let someone else read the correct answer, and you try to recite the exact question! This works out particularly well as a group exercise.

6. Once you have most of the questions "down cold," begin checking them off your list. Put a check beside each question you know perfectly. Then concentrate on those harder questions that may require some memorization and better understanding. Check them off as you are sure you know them.

7. Try a practice examination and see how well you do. Maybe you can get some ham buddies to make one up for you.

8. If you are planning to take the code test, build your speed to a minimum of 5 wpm. If you can pass the code test, you will pick up some exciting additional privileges on the worldwide ham bands.

EXAMINATION QUESTION POOL

Are you ready to prepare for your Technician Class no-code or Novice Class license or your Technician-Plus privileges? Here are the two question pools. Let's get started learning these questions and answers. Study hard, and good luck!

When you complete your study, head for the local accredited VE team to take your 2-part examination in one sitting. This is a good way to go, and gets everything out of the way all at once. You might even consider taking the Element 1A code test and end up a Technician-Plus operator.

Element 2 and Element 3A Question Pools

Element 2 has 350 questions
Element 3A has 295 questions

Written examinations will be prepared by selecting questions from these question pools as follows:

- ■ *Novice Class* applicants take:

  a 30-question examination from Element 2

- ■ *Technician No-Code* applicants take:

  a 30-question examination from Element 2

  and

  a 25-question examination from Element 3A

- ■ *Technician-Plus* applicants Upgrading from Novice take:

  a 25-question examination from Element 3A

QUESTION POOL SYLLABUS

The syllabus used for the development of the question pool is included at the beginning of each question pool as an aid in studying the subelements and topic groups. Review each syllabus before you start your study to gain an understanding of the question pool details.

Element 2 (Novice Class) Syllabus

Here is the syllabus used to develop the question pool:

**N1 – Commission's Rules
(10 exam questions – 10 groups)**
N1A Basis and purpose of amateur service and definitions
N1B Station/Operator license
N1C Novice control operator frequency privileges
N1D Novice eligibility, exam elements, mailing addresses, US call sign assignment and life of license
N1E Novice control operator emission privileges
N1F Transmitter power on Novice sub-bands and digital communications (limited to concepts only)
N1G Responsibility of licensee, control operator requirements
N1H Station identification, points of communication and operation, and business communications
N1I International and space communications, authorized and prohibited transmissions
N1J False signals or unidentified communications and malicious interference

**N2 – Operating Procedures
(2 exam questions – 2 groups)**
N2A Choosing a frequency for tune-up, operating or emergencies; understanding the Morse code; RST signal reports; Q signals; voice communications and phonetics
N2B Radio teleprinting; packet; repeater operating procedures; special operations

**N3 – Radio Wave Propagation
(1 exam question – 1 group)**
N3A Radio wave propagation, line of sight, ground wave, sky wave, sunspots and the sunspot cycle, reflection of VHF/UHF signals

**N4 – Amateur Radio Practices
(4 exam questions – 4 groups)**
N4A Unauthorized use prevention, lightning protection, and station grounding
N4B Radio frequency safety precautions, safety interlocks, antenna installation safety procedures
N4C SWR meaning and measurements
N4D RFI and its complications

**N5 – Electrical Principles
(4 exam questions – 4 groups)**
N5A Metric prefixes, i.e. pico, micro, milli, centi, kilo, mega, giga
N5B Concepts of current, voltage, conductor, insulator, resistance, and the measurement thereof
N5C Ohm's Law {any calculations will be kept to a very low level — no fractions or decimals} and the concepts of energy and power, and open and short circuits
N5D Concepts of frequency, including AC vs DC, frequency units, AF vs RF and wavelength

**N6 – Circuit Components
(2 exam questions – 2 groups)**
N6A Electrical function and/or schematic representation of resistor, switch, fuse, or battery
N6B Electrical function and/or schematic representation of a ground, antenna, transistor, or a triode vacuum tube

**N7 – Practical Circuits
(2 exam questions – 2 groups)**
N7A Functional layout of transmitter, transceiver, receiver, power supply, antenna, antenna switch, antenna feed line, impedance matching device, SWR meter
N7B Station layout and accessories for telegraphy, radiotelephone, radioteleprinter or packet

**N8 – Signals and Emission
(2 exam questions – 2 groups)**
N8A Emission types, key clicks, chirps or superimposed hum
N8B Harmonics and unwanted signals, equipment and adjustments to help reduce interference to others

**N9 – Antennas and Feed Lines
(3 exam questions – 3 groups)**
N9A Wavelength vs antenna length
N9B Yagi parts, concept of directional antennas, and safety near antennas
N9C Feed lines, baluns and polarization via element orientation

Subelement N1 – Commission's Rules	10 exam questions
	10 topic groups

Note: A §Part 97 reference is enclosed in brackets, e.g., [97], after each correct answer explanation in this subelement.

N1A Basis and purpose of amateur service and definitions

N1A01 What document contains the rules and regulations for the amateur service in the US?
- A. Part 97 of Title 47 CFR (Code of Federal Regulations)
- B. The Communications Act of 1934 (as amended)
- C. The Radio Amateur's Handbook
- D. The minutes of the International Telecommunication Union meetings

ANSWER A: A copy of §Part 97 of Title 47 of the Code of Federal Regulations should be in every amateur's library. [97]

N1A02 Who makes and enforces the rules and regulations of the amateur service in the US?
- A. The Congress of the United States
- B. The Federal Communications Commission (FCC)
- C. The Volunteer Examiner Coordinators (VECs)
- D. The Federal Bureau of Investigation (FBI)

ANSWER B: The Federal Communications Commission (FCC) makes and enforces all amateur radio rules in the United States. [97]

N1A03 Which three topics are part of the rules and regulations of the amateur service?
- A. Station operation standards, technical standards, emergency communications
- B. Notice of Violation, common operating procedures, antenna lengths
- C. Frequency band plans, repeater locations, Ohm's law
- D. Station construction standards, FCC approved radios, FCC approved antennas

ANSWER A: The FCC rules determine how we operate on the air, technical standards of our equipment, and emergency communications. The rules do not cover how to operate, repeater locations, nor do the rules cover how your station is actually constructed. Review the wrong answers in order to remember answer A as the correct answer. [97]

N1A04 Which of these topics is NOT part of the rules and regulations of the amateur service?
- A. Qualifying examination systems
- B. Technical standards
- C. Providing emergency communications
- D. Station construction standards

ANSWER D: Station construction standards are not addressed in the FCC rules and regulations; however, assemble your station carefully and safely, ensuring that everything is well-grounded. [97]

N1A05 What are three reasons that the amateur service exists?

A. To recognize the value of emergency communications, advance the radio art, and improve communication and technical skills
B. To learn about business communications, increase testing by trained technicians, and improve amateur communications
C. To preserve old radio techniques, maintain a pool of people familiar with early tube-type equipment, and improve tube radios
D. To improve patriotism, preserve nationalism, and promote world peace

ANSWER A: Memorize these principles. They truly sum up what ham radio is all about. Read over the incorrect answers, and see the hidden wording that makes them wrong. [97.1]

N1A06 What are two of the five purposes for the amateur service?

A. To protect historical radio data, and help the public understand radio history
B. To help foreign countries improve communication and technical skills, and encourage visits from foreign hams
C. To modernize radio schematic drawings, and increase the pool of electrical drafting people
D. To increase the number of trained radio operators and electronics experts, and improve international goodwill

ANSWER D: The Federal Communications Commission rules, Part 97, describes the amateur service as a radiocommunications service for the purpose of self-training, intercommunication and technical investigation, carried out by amateurs, that is, duly authorized persons interested in radio technique solely with a personal aim and without pecuniary interest. While it is true that ham operators are known for their emergency communications assistance in times of need, the amateur service is basically a hobby radio service, just for fun! [97.1]

N1A07 What is the definition of an amateur operator?

A. A person who has not received any training in radio operations
B. A person who has a written authorization to be the control operator of an amateur station
C. A person who has very little practice operating a radio station
D. A person who is in training to become the control operator of a radio station

ANSWER B: This will be you when you pass your 30-question Novice written theory exam and a 5 words-per-minute Morse code test before volunteer ham examiners. You will hold a valid ten-year license to operate an amateur station. The license will be issued by the Federal Communications Commission. [97.3a1]

N1A08 What is the definition of the amateur service?

A. A private radio service used for profit and public benefit
B. A public radio service for US citizens which requires no exam
C. A personal radio service used for self-training, communication, and technical studies
D. A private radio service used for self-training of radio announcers and technicians

ANSWER C: We sometimes call the amateur service "ham radio." What makes our hobby fun is experimenting and training ourselves in front of our ham radios. [97.3a4]

N1A09 What is the definition of an amateur station?
A. A station in a public radio service used for radiocommunications
B. A station using radiocommunications for a commercial purpose
C. A station using equipment for training new radiocommunications operators
D. A station in an Amateur Radio service used for radiocommunications

ANSWER D: Your ham station will consist of a radio device, at a particular location, that will be used for amateur communications. This station can go anywhere that you go, and might be mobile, a base, or a hand-held set. You also are allowed to choose any authorized frequency in any authorized band. You can change radio equipment type at anytime. [97.3a5]

N1A10 What is the definition of a control operator of an amateur station?
A. Anyone who operates the controls of the station
B. Anyone who is responsible for the station's equipment
C. Any licensed amateur operator who is responsible for the station's transmissions
D. The amateur operator with the highest class of license who is near the controls of the station

ANSWER C: This is a fancy name for you, the person who holds an amateur operator/primary station license. You, or any other ham you designate, are in control of all transmissions, and are responsible for the proper operation of the station. [97.3a11]

N1A11 What is a Volunteer Examiner (VE)?
A. An amateur who volunteers to check amateur teaching manuals
B. An amateur who volunteers to teach amateur classes
C. An amateur who volunteers to test others for amateur licenses
D. An amateur who volunteers to examine amateur station equipment

ANSWER C: Volunteer examiners (VEs) administer ham radio examinations from Novice through Extra Class. No longer do you take your test in front of an FCC employee. [97.513a]

N1B Station/Operator license

N1B01 Which one of these must you have an amateur license to do?
A. Transmit on public-service frequencies
B. Retransmit shortwave broadcasts
C. Repair broadcast station equipment
D. Transmit on amateur service frequencies

ANSWER D: You need an amateur license to transmit on amateur service frequencies. The license for public safety, shortwave, and repairing broadcast station equipment is different than the amateur license. [97.5a]

N1B02 What does an amateur license allow you to control?
A. A shortwave-broadcast station's transmissions
B. An amateur station's transmissions

C. Non-commercial FM broadcast transmissions

D. Any type of transmitter, as long as it is used for non-commercial transmissions

ANSWER B: Your station license authorizes you to use any type of Amateur Radio equipment at any particular location. You may operate your radio equipment almost anywhere in the United States and its possessions, except in commercial airplanes. But you don't have to operate your station at the address you list on your application as your current primary station location. [97.5a]

N1B03 What allows someone to operate an amateur station in the US?

A. An FCC operator's training permit for a licensed radio station

B. An FCC Form 610 together with a license examination fee

C. An FCC amateur operator/primary station license

D. An FCC Certificate of Successful Completion of Amateur Training

ANSWER C: You will receive your operator license and station call sign directly from the Federal Communications Commission. You will be the only one in the world with that call sign. It takes about 30 days after you pass your license exam to receive your call letters. Make several copies of your license, and always keep a copy of your operator license with you when you operate your equipment. [97.5a]

N1B04 Where does a US amateur license allow you to operate?

A. Anywhere in the world

B. Wherever the amateur service is regulated by the FCC

C. Within 50 km of your primary station location

D. Only at your primary station location

ANSWER B: Once you have a license, you may operate anywhere the FCC has jurisdiction. About the only exception is in a commercial aircraft. No operation is allowed without special permission. [97.5d]

N1B05 If you have a Novice license, how many transmitters may you control in your station at the same time?

A. Only one at a time

B. Only one at a time, except for emergency communications

C. Any number

D. Any number, as long as they are transmitting on different bands

ANSWER C: You may have any number of radios at your station. Your author has over 25 radios in his shack! [97.5e]

N1B06 What document must you keep at your amateur station?

A. A copy of your written authorization for an amateur station

B. A copy of the Rules and Regulations of the Amateur Service (Part 97)

C. A copy of the Amateur Radio Handbook for instant reference

D. A chart of the frequencies allowed for your class of license

ANSWER A: Keep the original of your license safe and sound at your station. Make plenty of copies for multiple control points. Author's quote: "I have copies of my license in my car, on the boat, and in my airplane." [97.5e]

N1B07 Which one of the following does not allow a person to control a US amateur station?

A. An operator/primary station license from the FCC

B. A reciprocal permit for alien amateur licensee from the FCC

C. An amateur service license from any government which is a member of the European Community (EC)

D. An amateur service license from the Government of Canada, if it is held by a Canadian citizen

ANSWER C: European amateur radio operators are not allowed to control a U.S. amateur station with their foreign license. To operate a U.S. amateur station, you need an FCC amateur operator/primary station license, or an FCC alien amateur license, or a Canadian license if it is held by a Canadian citizen. [97.7]

N1B08 What are the five US amateur operator license classes?

A. Novice, Communicator, General, Advanced, Amateur Extra

B. Novice, Technician, General, Advanced, Expert

C. Novice, Communicator, General, Amateur, Extra

D. Novice, Technician, General, Advanced, Amateur Extra

ANSWER D: Memorize the order of amateur operator licenses. Remember, all ham licenses are additive—you can't skip over any license category to get to the top one. [97.9a]

N1B09 What does the FCC consider to be the first two classes of US amateur operator licenses (one of which most new amateurs initially hold)?

A. Novice and Technician

B. CB and Communicator

C. Novice and General

D. CB and Novice

ANSWER A: The Novice license is the easiest one to obtain. There is no age limit, and kids as young as five years old have passed the test. [97.9]

N1B10 What must you have with you when you are the control operator of an amateur station?

A. A copy of the Rules and Regulations of the Amateur Service (Part 97)

B. The original or a photocopy of your amateur license

C. A list of countries which allow third-party communications from the US

D. A chart of the frequencies allowed for your class of license

ANSWER B: If you are visiting another ham's station, and you operate the equipment with you acting as a control operator, both you and the other ham are jointly responsible for the proper operation of the station. [97.9]

N1B11 Which US amateur license has no Morse code requirements?

A. Amateur Extra

B. Advanced

C. General

D. Technician

ANSWER D: The Technician Class no-code license does not require a Morse code test. [97.501d]

N1C Novice control operator frequency privileges

N1C01 What are the frequency limits of the 80-meter Novice band?
   A. 3500 - 4000 kHz
   B. 3675 - 3725 kHz
   C. 7100 - 7150 kHz
   D. 7000 - 7300 kHz

ANSWER B: Memorize this frequency range, and use the frequency chart in this book to help you visualize where this band is on the radio dial. The 80-meter wavelength band is great for Novices at night for long-range, CW contacts. [97.301e]

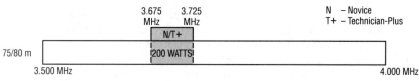

75/80-Meter Wavelength Band Privileges

N1C02 What are the frequency limits of the 40-meter Novice band (ITU Region 2)?
   A. 3500 - 4000 kHz
   B. 3700 - 3750 kHz
   C. 7100 - 7150 kHz
   D. 7000 - 7300 kHz

ANSWER C: This is the 40-meter wavelength band, so memorize the frequencies. The 40-meter band is great at nighttime for long-range, Novice, CW contacts. During the day, CW contacts may be made up to 400 miles away. [97.301e]

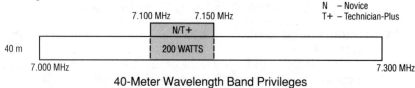

40-Meter Wavelength Band Privileges

N1C03 What are the frequency limits of the 15-meter Novice band?
   A. 21.100 - 21.200 MHz
   B. 21.000 - 21.450 MHz
   C. 28.000 - 29.700 MHz
   D. 28.100 - 28.200 MHz

ANSWER A: Here's where the CW fun begins for the Novice on 15 meters— during the day and evening. You can usually exceed 3,000 miles with a telegraph key! [97.301e]

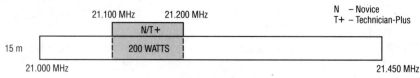

15-Meter Wavelength Band Privileges

N1C04 What are the frequency limits of the 10-meter Novice band?
A. 28.000 - 28.500 MHz
B. 28.100 - 29.500 MHz
C. 28.100 - 28.500 MHz
D. 29.100 - 29.500 MHz

ANSWER C: 10 meters is a "daytime" band. It's divided in half. 28.3 to 28.5 MHz is where most Novices hang out for long-range voice contacts. 28.1 to 28.2 MHz is a great place for CW. 28.2 to 28.3 MHz is where automatic beacons transmit propagation CW signals. [97.301e]

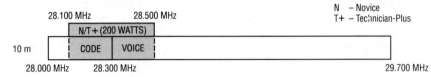

10-Meter Wavelength Band Privileges

N1C05 What are the frequency limits of the 1.25-meter Novice band (ITU Region 2)?
A. 225.0 - 230.5 MHz
B. 222.1 - 223.91 MHz
C. 224.1 - 225.1 MHz
D. 222 - 225 MHz

ANSWER B: Plenty of repeater activity here. Memorize the 222-MHz band frequencies. See figure at question N1C06. [97.301f]

1.25-Meter Wavelength Band Privileges

N1C06 What are the frequency limits of the 23-centimeter Novice band?
A. 1260 - 1270 MHz
B. 1240 - 1300 MHz
C. 1270 - 1295 MHz
D. 1240 - 1246 MHz

ANSWER C: This band is just becoming popular. Your author uses it for repeater operation throughout Southern California. It's an exceptional band for small hand-held transceivers and mobile units in big cities where signals are reflected by many buildings. Memorize these frequencies. [97.301f]

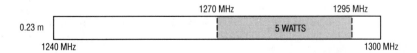

0.23-Meter (23-Centimeters) Wavelength Band Privileges

N1C07 If you are operating on 3700 kHz, in what amateur band are you operating?
  A. 80 meters
  B. 40 meters
  C. 15 meters
  D. 10 meters

ANSWER A: Look at the frequency privileges in the authorized frequency bands chart in the Appendix of this book, and memorize the MHz frequencies as they relate to the band in meters. To convert MHz to meters, first convert kHz (3700) to MHz (3.700). Now divide this into 300, and presto, you end up with 80 meters. The answer is rounded to 80 because the wavelength for a band is an average number broadly covering all the frequencies in the band. The meter band is the wavelength of the operating frequency. Wavelength is found by the equation:

$$\lambda(\text{Wavelength in meters}) = \frac{300}{f(\text{MHz})}$$

which says that the wavelength in meters is equal to 300 divided by the frequency in megahertz. With a calculator, the keystrokes are: CLEAR 300 ÷ 3.7 =. The answer is 81.08 rounded to 80 meters. [97.301e]

Converting Frequency to Wavelength	Converting Wavelength to Frequency
To find wavelength (λ) in meters, if you know frequency (f) in megahertz (MHz),	To find frequency (f) in megahertz (MHz), if you know wavelength (λ) in meters,
Solve:	**Solve:**
$\lambda(\text{meters}) = \dfrac{300}{f(\text{MHz})}$	$f(\text{MHz}) = \dfrac{300}{\lambda(\text{meters})}$

Conversions Between Wavelength and Frequency

N1C08 If you are operating on 7125 kHz, in what amateur band are you operating?
  A. 80 meters
  B. 40 meters
  C. 15 meters
  D. 10 meters

ANSWER B: Novices in North and South America may operate their transmitters using telegraphy only between 7100 and 7150 kHz. The 7125 kHz frequency is located in the 40-meter ham band. Remember 7125 kHz is 7.125 MHz. Use the equation in question N1C07. The calculator keystrokes are: CLEAR 300 ÷ 7.125 =. The answer, 42.1, is rounded to 40 meters. [97.301e]

N1C09 If you are operating on 21.150 MHz, in what amateur band are you operating?
  A. 80 meters
  B. 40 meters
  C. 15 meters
  D. 10 meters

ANSWER C: Use the equation in question N1C07. Convert 21,150 kHz into MHz. That's right, 21.150 MHz. Now divide this into 300, and you end up with 14.2 meters. Round this up to 15 meters to match the closest answer choice. The calculator keystrokes are: CLEAR 300 ÷ 21.15 =. [97.301e]

N1C10 If you are operating on 28.150 MHz, in what amateur band are you operating?
    A. 80 meters
    B. 40 meters
    C. 15 meters
    D. 10 meters

ANSWER D: Use the equation in question N1C07. 28,150 kHz is 28.150 MHz. Don't knock yourself out with precise division—just run 28 MHz into 300, and that comes out close to 10 meters. That's the answer. The calculator keystrokes are: CLEAR 300 ÷ 28.15 =. The 10.7 answer is rounded down to 10. [97.301e]

N1C11 If you are operating on 223 MHz, in what amateur band are you operating?
    A. 15 meters
    B. 10 meters
    C. 2 meters
    D. 1.25 meters

ANSWER D: The 223 MHz band is more commonly called 1.25 meters, or sometimes the 1-1/4-meter band. Check this by using the equation in question N1C07. The calculator keystrokes are: CLEAR 300 ÷ 223 =. The closest answer choice to 1.35 meters is 1.25 meters. [97.301f]

N1D Novice eligibility, exam elements, mailing addresses, US call-sign assignment and life of license

N1D01 Who can become an amateur licensee in the US?
    A. Anyone except a representative of a foreign government
    B. Only a citizen of the United States
    C. Anyone except an employee of the US government
    D. Anyone

ANSWER A: There is no nationality requirement to become an amateur operator in the United States, but representatives of foreign governments are ineligible. Examinations and code tests are in English, so the applicant must be able to speak and understand English. [97.5d1]

N1D02 What age must you be to hold an amateur license?
    A. 14 years or older
    B. 18 years or older
    C. 70 years or younger
    D. There are no age limits

ANSWER D: There are amateur radio operators that are as young as 5 years old! Also, most kids love the Morse code.

N1D03 What minimum examinations must you pass for a Novice amateur license?

    A. A written exam, Element 1(A); and a 5-WPM code exam, Element 2(A)

    B. A 5-WPM code exam, Element 1(A); and a written exam, Element 3(A)

    C. A 5-WPM code exam, Element 1(A); and a written exam, Element 2

    D. A written exam, Element 2; and a 5-WPM code exam, Element 4

ANSWER C: Element 1A is the 5-wpm Morse code test. Element 2 is the Novice written examination. [97.501e]

N1D04 Why must an amateur operator have a current US Postal mailing address?

    A. So the FCC has a record of the location of each amateur station

    B. To follow the FCC rules and so the licensee can receive mail from the FCC

    C. So the FCC can send license-renewal notices

    D. So the FCC can publish a call-sign directory

ANSWER B: When you fill out the FCC Form 610 (one is in the back of this book) to apply for your amateur operator/primary station license, it asks for your mailing address. Where do you want to get the license? At your home? At your office? At a friend's house? It's a hassle to change your mailing address, so make it some spot that you plan to get your mail for the next few years. [97.21]

N1D05 What must you do to replace your license if it is lost, mutilated or destroyed?

    A. Nothing; no replacement is needed

    B. Send a change of address to the FCC using a current FCC Form 610.

    C. Retake all examination elements for your license.

    D. Request a new one from the FCC, explaining what happened to the original.

ANSWER D: If your dog ate your license, or you simply lost it, write the Federal Communications Commission, 1270 Fairfield Road, Gettysburg, Pennsylvania 17325, explain how the license was lost or destroyed, and ask them for a replacement. Keep a copy of this letter with your station records until your replacement license arrives in about 70 days. [97.27]

N1D06 What must you do to notify the FCC if your mailing address changes?

    A. Fill out an FCC Form 610 using your new address, attach a copy of your license, and mail it to your local FCC Field Office.

    B. Fill out an FCC Form 610 using your new address, attach a copy of your license, and mail it to the FCC office in Gettysburg, PA.

    C. Call your local FCC Field Office and give them your new address over the phone.

    D. Call the FCC office in Gettysburg, PA, and give them your new address over the phone.

ANSWER B: You will need to fill out an FCC Form 610 if you have changed your address. Send a copy of your license, not the original, to the FCC in Gettysburg, Pennsylvania. NEVER EVER give up your original license. Copies are fine. [97.19]

N1D07 Which of the following call signs is a valid US amateur call?
    A. UA4HAK
    B. KBL7766
    C. KA9OLS
    D. BY7HY

ANSWER C: There are never more than two letters preceding the number in a ham call sign, followed by another letter, or two, or three. Answer B has too many letters, and ends up with numbers—so it's incorrect. And since Answers A and D don't begin with A, K, N, or W, they also are wrong.

N1D08 What letters must be used for the first letter in US amateur call signs?
    A. K, N, U and W
    B. A, K, N and W
    C. A, B, C and D
    D. A, N, V and W

ANSWER B: In the United States, all ham call signs begin with A, K, N, or W. This is because, as an aid to enforcement, all transmitting stations are required to identify themselves at intervals when they are in operation. Radio does not respect national boundaries. By international agreement, the first characters of the call sign indicate the country in which the station is authorized to operate. The only prefixes allocated to United States amateur stations are: AA-AL, KA-KZ, NA-NZ, and WA-WZ.

N1D09 What numbers are normally used in US amateur call signs?
    A. Any two-digit number, 10 through 99
    B. Any two-digit number, 22 through 45
    C. A single digit, 1 though 9
    D. A single digit, 0 through 9

ANSWER D: You can get a good idea where someone is by the number in their call sign—unless they moved to another state and didn't change their call sign. You can buy colorful charts that illustrate the geographic area assigned to Amateur Radio call sign numbers.

U.S. Call Sign Areas

N1D10 For how many years is an amateur license normally issued?

A. 2
B. 5
C. 10
D. 15

ANSWER C: Hams now have a 10-year period for their license before it needs to be renewed. There is a 2-year grace period if you forget to renew—but you can't operate after your license has expired. After the 2-year license expiration grace period, your license is lost, and you have to start all over again. Don't forget to renew! [97.23]

N1D11 How soon before your license expires should you send the FCC a completed 610 for a renewal?

A. 60 to 90 days
B. Within 21 days of the expiration date
C. 6 to 9 months
D. 6 months to a year

ANSWER A: You should fill out a Form 610 at least 90 days in advance for a license renewal. The FCC does not send you a renewal reminder. If you forget to renew your license, and the 2-year grace period expires, you will have to start all over again. [97.19c]

N1E Novice control operator emission privileges

N1E01 What emission types are Novice control operators allowed to use in the 80-meter band?

A. CW only
B. Data only
C. RTTY only
D. Phone only

ANSWER A: CW only means Morse code. You may operate telegraphy on 3675-3725 kHz, your Novice 80-meter wavelength band privileges. A1A is a code commonly used in early Amateur Radio to designate CW emissions. [97.305/.307f9]

> **CW** – International Morse code telegraphy emissions.
> **Data** – Telemetry, telecommand and computer communications emissions.
> **RTTY** (Radioteletype) – Narrow-band, direct-printing telegraphy emissions.
> **Phone** – Speech and other sound emissions.

Emission Definitions

N1E02 What emission types are Novice control operators allowed to use in the 40-meter band?

A. CW only
B. Data only
C. RTTY only
D. Phone only

ANSWER A: Novice hams may use Morse code (CW, A1A) on 40 meters, 7100-7150 kHz. [97.305/.307f9]

N1E03 What emission types are Novice control operators allowed to use in the 15-meter band?
A. CW only
B. Data only
C. RTTY only
D. Phone only
ANSWER A: Morse code only (CW, A1A), from 21,100-21,200 kHz. [97.305/.307f9]

N1E04 What emission types are Novice control operators allowed to use from 3675 to 3725 kHz?
A. Phone only
B. Image only
C. Data only
D. CW only
ANSWER D: Novices may use Morse code only on 80 meters. [97.305/.307f9]

N1E05 What emission types are Novice control operators allowed to use from 7100 to 7150 kHz in ITU Region 2?
A. CW and data
B. Phone
C. Data only
D. CW only
ANSWER D: Your Novice station can transmit continuous wave (CW) or A1A only on 40 meters. [97.305/.307f9]

N1E06 What emission types are Novice control operators allowed to use on frequencies from 21.1 to 21.2 MHz?
A. CW and data
B. CW and phone
C. Data only
D. CW only
ANSWER D: On 15 meters, an exceptional long-range band for daytime and evening contacts, Novices are allowed only CW (A1A), Morse code. [97.305/.307f9]

N1E07 What emission types are Novice control operators allowed to use on frequencies from 28.1 to 28.3 MHz?
A. All authorized amateur emission privileges
B. Data or phone
C. CW, RTTY and data
D. CW and phone
ANSWER C: It is okay for Novices to operate digital computers between 28.1 and 28.3. Besides data, you can also operate CW (A1A), Morse code. [97.305]

N1E08 What emission types are Novice control operators allowed to use on frequencies from 28.3 to 28.5 MHz?
A. All authorized amateur emission privileges
B. CW and data
C. CW and single-sideband phone
D. Data and phone

ANSWER C: Besides Morse code on the voice portion of the Novice band on 10 meters, you may also operate that high-frequency voice mode called single sideband. [97.305/.307f10]

N1E09 What emission types are Novice control operators allowed to use on the amateur 1.25-meter band in ITU Region 2?
    A. CW and phone
    B. CW and data
    C. Data and phone
    D. All amateur emission privileges authorized for use on the band
ANSWER D: Within your Novice privileges at 222.1 to 223.91, you may transmit any emission of your choice. Your repeated output may also legally exceed your Novice band limits. [97.305]

N1E10 What emission types are Novice control operators allowed to use on the amateur 23-centimeter band?
    A. Data and phone
    B. CW and data
    C. CW and phone
    D. All amateur emission privileges authorized for use on the band
ANSWER D: Within your privileges from 1270 to 1295 MHz, you may operate any emission privileges—including television and all forms and speeds of packet. [97.305]

N1E11 On what HF frequencies may Novice control operators use single-sideband (SSB) phone?
    A. 3700 - 3750 kHz
    B. 7100 - 7150 kHz
    C. 21100 - 21200 kHz
    D. 28300 - 28500 kHz
ANSWER D: There is plenty of long-range excitement on 28.3 to 28.5 MHz, 10 meters! [97.305/.307f10]

N1E12 On what frequencies in ITU Region 2 may Novice control operators use FM phone?
    A. 28.3 - 28.5 MHz
    B. 144.0 - 148.0 MHz
    C. 222.1 - 223.91 MHz
    D. 1240 - 1270 MHz
ANSWER C: Memorize these frequencies. They are your transmit privileges on the amateur service 222-MHz band. And good news—as long as you transmit within your own band limits, you may operate through 222-MHz repeaters that are located just outside Novice band limits—legally! So if someone asks you to go to a repeater on a frequency just outside of band limits for Novice, chances are the input is within Novice limits, and that's perfectly legal. Novice operators may soon have all of 222-225 MHz, pending new rule making. [97.305]

N1E13 On what frequencies in the 10-meter band may Novice control operators use RTTY?
    A. 28.0 - 28.3 MHz
    B. 28.1 - 28.3 MHz

C. 28.0 - 29.3 MHz

D. 29.1 - 29.3 MHz

ANSWER B: RTTY stands for radio teletypewriter, and you can transmit and receive radio teletype messages as well as Morse code from 28.1 to 28.3 MHz. [97.301e/.305]

N1E14 On what frequencies in the 10-meter band may Novice control operators use data emissions?

A. 28.0 - 28.3 MHz

B. 28.1 - 28.3 MHz

C. 28.0 - 29.3 MHz

D. 29.1 - 29.3 MHz

ANSWER B: The frequencies between 28.1 and 28.3 are filled with data transmissions, including packet and RTTY. Remember, a Technician-Plus Class operator (plus meaning code) gains Novice control operator frequency privileges on the 10-meter band, plus CW on 15 meters, 40 meters, and 80 meters. [97.301e/.305]

N1F Transmitter power on Novice sub-bands and digital communications [limited to concepts only]

N1F01 What amount of transmitter power must amateur stations use at all times?

A. 25 watts PEP output

B. 250 watts PEP output

C. 1500 watts PEP output

D. The minimum legal power necessary to communicate

ANSWER D: Be careful of this one—the answer is not a numerical one, but rather a philosophical one. Always run the minimum amount of power to make contact with another station. [97.313a]

N1F02 What is the most transmitter power an amateur station may use on 3700 kHz?

A. 5 watts PEP output

B. 25 watts PEP output

C. 200 watts PEP output

D. 1500 watts PEP output

ANSWER C: Everyone, including Extra Class operators, must stay below 200 watts PEP output. [97.313c1]

N1F03 What is the most transmitter power an amateur station may use on 7125 kHz?

A. 5 watts PEP output

B. 25 watts PEP output

C. 200 watts PEP output

D. 1500 watts PEP output

ANSWER C: Here is another Novice CW sub-band, so every ham in the U.S. may not run more than 200 watts PEP output. [97.313c1]

N1F04 What is the most transmitter power an amateur station may use on 21.125 MHz?
A. 5 watts PEP output
B. 25 watts PEP output
C. 200 watts PEP output
D. 1500 watts PEP output
ANSWER C: And yet another Novice sub-band, where under 200 watts PEP output is the legal limit. [97.313c1]

N1F05 What is the most transmitter power a Novice station may use on 28.125 MHz?
A. 5 watts PEP output
B. 25 watts PEP output
C. 200 watts PEP output
D. 1500 watts PEP output
ANSWER C: As a Novice operator, you are restricted to 200 watts PEP output, or less, when operating on the 10-meter band. [97.313c2]

N1F06 What is the most transmitter power a Novice station may use on the 10-meter band?
A. 5 watts PEP output
B. 25 watts PEP output
C. 200 watts PEP output
D. 1500 watts PEP output
ANSWER C: Although 200 watts output on 10 meters is allowed, try to keep your power output to a minimum. This band could cause second harmonic interference to your neighbors on television Channel 2. [97.313c2]

N1F07 What is the most transmitter power a Novice station may use on the 1.25-meter band?
A. 5 watts PEP output
B. 25 watts PEP output
C. 200 watts PEP output
D. 1500 watts PEP output
ANSWER B: On this VHF band, your power output is limited to 25 watts by FCC order. Don't worry about this limitation—almost everybody else on this band runs no more than 25 watts of power anyway! That's plenty of power to work any repeater you can hear. [97.313d]

N1F08 What is the most transmitter power a Novice station may use on the 23-centimeter band?
A. 5 watts PEP output
B. 25 watts PEP output
C. 200 watts PEP output
D. 1500 watts PEP output
ANSWER A: The 1270-MHz band is in the microwave region. You are approaching frequencies that will cook a turkey in a microwave oven. The FCC limits your power output to 5 watts to keep you from accidentally frying yourself in this region. Most other hams are also operating at the 5-watt level to reduce their exposure to unnecessary amounts of RF radiation. [97.313e]

N1F09 On which bands may a Novice station use up to 200 watts PEP output power?
A. 80, 40, 15, and 10 meters
B. 80, 40, 20, and 10 meters
C. 1.25 meters
D. 23 centimeters
ANSWER A: Here they want to know if you actually understand the maximum power level in watts. Novice bands on 80, 40, 15, and 10 meters are protected from high-power operation by any class of operator. Everyone within your Novice band limits must reduce power down to 200 watts PEP output or less. This means you won't have to battle a kilowatt station within your CW band privileges. [97.313c]

N1F10 On which bands must a Novice station use no more than 25 watts PEP output power?
A. 80, 40, 15, and 10 meters
B. 80, 40, 20, and 10 meters
C. 1.25 meters
D. 23 centimeters
ANSWER C: Novice stations may transmit no more than 25 watts output on the 222-225 MHz band. This is for their personal safety. [97.313d]

N1F11 On which bands must a Novice station use no more than 5 watts PEP output power?
A. 80, 40, 15, and 10 meters
B. 80, 40, 20, and 10 meters
C. 1.25 meters
D. 23 centimeters
ANSWER D: Up on 23 cm, emission types are microwave, and only 5 watts of output power are allowed for Novice operators. This is for the personal safety of Novice operators. [97.313e]

N1G Responsibility of licensee, control operator requirements

N1G01 If you allow another amateur to be responsible for the transmissions from your station, what is the other operator called?
A. An auxiliary operator
B. The operations coordinator
C. A third-party operator
D. A control operator
ANSWER D: If you are designated a control operator, you have a big responsibility ahead of you. Always remember that you are responsible for the proper operation of the station no matter who operates the key or talks over the mike. [97.3a11]

N1G02 Who is responsible for the proper operation of an amateur station?
A. Only the control operator
B. Only the station licensee
C. Both the control operator and the station licensee
D. The person who owns the station equipment

ANSWER C: If you are visiting another ham's station, and you operate the equipment with you acting as a control operator, both you and the other ham are jointly responsible for the proper operation of the station. [97.103a]

N1G03 If you transmit from another amateur's station, who is responsible for its proper operation?
  A. Both of you
  B. The other amateur (the station licensee)
  C. You, the control operator
  D. The station licensee, unless the station records show that you were the control operator at the time

ANSWER A: As a control operator, both you and the station licensee are responsible for the transmissions from that station. [97.103a]

N1G04 What is your responsibility as a station licensee?
  A. You must allow another amateur to operate your station upon request.
  B. You must be present whenever the station is operated.
  C. You must notify the FCC if another amateur acts as the control operator.
  D. You are responsible for the proper operation of the station in accordance with the FCC rules.

ANSWER D: When a fellow ham is using your equipment and station call sign, remember your responsibility of making sure everything is legal. [97.103a]

N1G05 Who may be the control operator of an amateur station?
  A. Any person over 21 years of age
  B. Any person over 21 years of age with a General class license or higher
  C. Any licensed amateur chosen by the station licensee
  D. Any licensed amateur with a Technician class license or higher

ANSWER C: As a Novice Class operator, you may only assume the control operator responsibility as your Novice license permits. If you are visiting another station operated by a General Class operator on General Class frequencies, your Novice Class license does not allow you to assume control operator functions. [97.103b]

N1G06 If another amateur transmits from your station, which of these is NOT true?
  A. You must first give permission for the other amateur to use your station.
  B. You must keep the call sign of the other amateur, together with the time and date of transmissions, in your station records.
  C. The FCC will think that you are the station's control operator unless your station records show that you were not.
  D. Both of you are equally responsible for the proper operation of the station.

ANSWER B: It's perfectly okay to let a fellow ham talk over your station. You do not need to keep a record of their call sign. In fact, it is no longer necessary to maintain a log of any amateur transmission; however, a record in a logbook is strongly recommended. Fill out log entries each time you use your station. [97.103]

N1G07 If you let another amateur with a higher class license than yours control your station, what operating privileges are allowed?

A. Any privileges allowed by the higher license
B. Only the privileges allowed by your license
C. All the emission privileges of the higher license, but only the frequency privileges of your license
D. All the frequency privileges of the higher license, but only the emission privileges of your license

ANSWER A: If a friend with a higher class license comes over and operates your equipment, he or she may transmit with the privileges allowed by his or her higher class license. Just because he or she is using your equipment doesn't change his or her privileges. [97.105b]

N1G08 If you are the control operator at the station of another amateur who has a higher class license than yours, what operating privileges are you allowed?

A. Any privileges allowed by the higher license
B. Only the privileges allowed by your license
C. All the emission privileges of the higher license, but only the frequency privileges of your license
D. All the frequency privileges of the higher license, but only the emission privileges of your license

ANSWER B: If you operate another amateur's equipment, you may operate only with the privileges allowed by your license, even though the equipment is owned by another ham with more privileges. [97.105b]

N1G09 When must an amateur station have a control operator?

A. Only when training another amateur
B. Whenever the station receiver is operated
C. Whenever the station is transmitting
D. A control operator is not needed

ANSWER C: Every ham station must have a responsible control operator unless it's an automated repeater station under "automatic control." The FCC defines a control operator as: "An amateur operator designated by the licensee of an amateur station to also be responsible for the emissions from that station." Automatic control means the use of devices or procedures for control without the control operator being present at the control point when the station is transmitting. [97.7]

N1G10 When a Novice station is transmitting, where must its control operator be?

A. At the station's control point
B. Anywhere in the same building as the transmitter
C. At the station's entrance, to control entry to the room
D. Anywhere within 50 km of the station location

ANSWER A: If you let another ham use your station, your control operator responsibilities require you to stay in the room, right at the radio equipment, supervising the communications. [97.109b]

N1G11 Why can't unlicensed persons in your family transmit using your amateur station if they are alone with your equipment?
- A. They must not use your equipment without your permission
- B. They must be licensed before they are allowed to be control operators
- C. They must first know how to use the right abbreviations and Q signals
- D. They must first know the right frequencies and emissions for transmitting

ANSWER B: In order to be a control operator for an amateur radio station, your family member must be properly licensed. [97.109b]

N1H Station identification, points of communication and operation, and business communications

N1H01 When may you operate your amateur station somewhere in the US besides the location listed on your license?
- A. Only during times of emergency
- B. Only after giving proper notice to the FCC
- C. During an emergency or an FCC-approved emergency practice
- D. Whenever you want to

ANSWER D: Just because your license has a permanent station location on it, don't think for a second that this is the only place you may operate your ham set. You can operate it anywhere in the U.S. and its territories and possessions without notifying the FCC. However, using your portable ham set on a commercial airplane is taboo. Using your ham set on cruise ships requires the permission of the captain. [97.5a]

N1H02 With which non-amateur stations is a US amateur station allowed to communicate?
- A. No non-amateur stations
- B. All non-amateur stations
- C. Only those authorized by the FCC
- D. Only those who use international Morse code

ANSWER C: In some very rare cases, the FCC may allow you to talk to a non-ham station—but usually only in an emergency. [97.111]

N1H03 When are communications for business allowed in the amateur service?
- A. Only if they are for the safety of human life or immediate protection of property.
- B. There are no rules against business communications.
- C. No business communications are ever allowed.
- D. Business communications are allowed between the hours of 9 AM and 5 PM, weekdays.

ANSWER A: Your local ham instructor has probably drilled into your memory that business communications are taboo on the ham bands. That's correct. However, during times of emergency, anything goes, including business. If you are providing emergency communications for a fire line, it's perfectly legal to use your hand-held to dial up an auto patch and order emergency food rations for the firefighters in action. "Let's see, I would like 2,000 Big Macs..." [97.113a]

N1H04 Which of the following CANNOT be discussed on an amateur club net?
 A. Business planning
 B. Recreation planning
 C. Code practice planning
 D. Emergency planning
ANSWER A: Until the rules are changed on the subject of business communications, amateur radio clubs are not allowed to conduct business regarding the club on the air. Other things like recreation, code practice, and emergency planning are perfectly okay. [97.113a]

N1H05 If you wanted to join a radio club, would you be allowed to send a message to them via amateur radio requesting an application?
 A. Yes, if the club is a not-for-profit organization
 B. No. This would facilitate the commercial affairs of the club.
 C. Yes, but only during normal business hours, between 9 AM and 5 PM, weekdays
 D. Yes, since there are no rules against business communications in the amateur service
ANSWER B: Using a ham set to handle any service from a company, or anything else of a business nature for you or the party you are calling, is not legal. [97.113a]

N1H06 How often must an amateur station be identified?
 A. At the beginning of a contact and at least every ten minutes after that
 B. At least once during each transmission
 C. At least every ten minutes during and at the end of a contact
 D. At the beginning and end of each transmission
ANSWER C: Give your call letters regularly. Remember, even though the law doesn't require that you give them at the beginning of the transmission, it makes good sense to start out with your call letters. [97.119a]

N1H07 What do you transmit to identify your amateur station?
 A. Your "handle"
 B. Your call sign
 C. Your first name and your location
 D. Your full name
ANSWER B: Use your own call sign—not someone else's call sign—when operating from your own station. [97.119a]

N1H08 What identification, if any, is required when two amateur stations begin communications?
 A. No identification is required.
 B. One of the stations must give both stations' call signs.
 C. Each station must transmit its own call sign.
 D. Both stations must transmit both call signs.
ANSWER A: Isn't that strange! By FCC law, you can actually communicate for nearly 10 minutes without identifying. You must identify at the ten minute point—or at the end of the communication, whichever comes first. Most hams identify when signing on a specific frequency so the other person will know who they are talking to. It is a good idea. [97.119a]

N1H09 What identification, if any, is required when two amateur stations end communications?

A. No identification is required
B. One of the stations must transmit both stations' call signs
C. Each station must transmit its own call sign
D. Both stations must transmit both call signs

ANSWER C: When you sign off, always end with your call sign. The other operator will always end with his or her call sign. [97.119a]

N1H10 Besides normal identification, what else must a US station do when sending third-party communications internationally?

A. The US station must transmit its own call sign at the beginning of each communication, and at least every ten minutes after that.
B. The US station must transmit both call signs at the end of each communication.
C. The US station must transmit its own call sign at the beginning of each communication, and at least every five minutes after that.
D. Each station must transmit its own call sign at the end of each communication, and at least every five minutes after that.

ANSWER B: It's common for U.S. amateur operators to handle third-party communications with those countries with whom we have a third-party agreement. The U.S. station must transmit both their call sign and the call sign of the foreign operator at the end of each communication. [97.115c]

N1H11 What is the longest period of time an amateur station can operate without transmitting its call sign?

A. 5 minutes
B. 10 minutes
C. 15 minutes
D. 20 minutes

ANSWER B: Some hams use a 10-minute timer to remind them to give their call signs. §Part 97 stipulates: "Each amateur shall give its call sign at the end of each communication, and every ten minutes or less during a communication." [97.119a]

N1I International and space communications, authorized and prohibited transmissions

---

N1I01 What is the definition of third-party communications?

A. A message sent between two amateur stations for someone else
B. Public service communications for a political party
C. Any messages sent by amateur stations
D. A three-minute transmission to another amateur

ANSWER A: Did you know you can let other people talk over your ham set who might not be ham radio operators? That's right, but you must stay right at the microphone to act as a "control operator" and make sure that they abide by the rules. When you link your ham radio into the telephone service, the people you can call would be considered "third-party." [97.3a39]

N1I02 When are you allowed to communicate with an amateur in a foreign country?

A. Only when the foreign amateur uses English
B. Only when you have permission from the FCC
C. Only when a third-party agreement exists between the US and the foreign country
D. At any time, unless it is not allowed by either government

ANSWER D: We may speak with every amateur operator in the world. Most foreign hams speak English as the common ham radio language. We are not prohibited from talking with any foreign ham radio operator at this time. [97.111a1]

N1I03 What is an amateur space station?
A. An amateur station operated on an unused frequency
B. An amateur station awaiting its new call letters from the FCC
C. An amateur station located more than 50 kilometers above the Earth's surface
D. An amateur station that communicates with Space Shuttles

ANSWER C: Special rules pertain to amateur operation in a space station. A space station is considered any amateur station located more than 50 kilometers above the earth's surface. Remote control of a model aircraft at 1000 feet is not considered a space station. [97.3a36]

N1I04 Who may be the licensee of an amateur space station?
A. An amateur holding an Amateur Extra Class operator license
B. Any licensed amateur operator
C. Anyone designated by the commander of the spacecraft
D. No one unless specifically authorized by the government

ANSWER B: It doesn't take any special grade of amateur operator license to be the licensee of an amateur space station. Years ago, only Extra Class amateur operators could go into space. Now, any licensed amateur operator may receive space station authorization. Shall we beam you up? [New 97.207a per FCC 92-310]

N1I05 When may someone be paid to transmit messages from an amateur station?
A. Only if he or she works for a public service agency such as the Red Cross
B. Under no circumstances
C. Only if he or she reports all such payments to the IRS
D. Only if he or she works for a club station and special requirements are met

ANSWER D: Normally you can't get paid to operate a ham station. However, there is one exception. If you are employed by an organization that transmits ham radio news bulletins and CW practice, it is permissible to draw wages for your work. [97.113b]

N1I06 When is an amateur allowed to broadcast information to the general public?
A. Never
B. Only when the operator is being paid
C. Only when broadcasts last less than 1 hour
D. Only when broadcasts last longer than 15 minutes

ANSWER A: News bulletins broadcast over the ham radio airwaves must relate solely to Amateur Radio matters or be of interest to [97.305] amateur operators who tune in, not to the general public. [97.113c]

N1I07 When is an amateur station permitted to transmit music?
 A. Never
 B. Only if the music played produces no spurious emissions
 C. Only if it is used to jam an illegal transmission
 D. Only if it is above 1280 MHz
ANSWER A: Don't sing "Happy Birthday" to a friend over ham radio. Music is not allowed. [97.113d]

N1I08 When is the use of codes or ciphers allowed to hide the meaning of an amateur message?
 A. Only during contests
 B. Only during nationally declared emergencies
 C. Never, except when special requirements are met
 D. Only on frequencies above 1280 MHz
ANSWER C: Secret codes are not allowed. It's even considered poor practice to use police-type "ten codes" on the air. [97.113d]

N1I09 What is a "third-party" in amateur communications?
 A. An amateur station that breaks in to talk
 B. A person who is sent a message by amateur communications other than a control operator who handles the message
 C. A shortwave listener who monitors amateur communications
 D. An unlicensed control operator
ANSWER B: When third-party traffic is taking place, never leave the room. You must supervise the conversation at all times to ensure that the rules concerning illegal messages are being abided by. All messages involving "business interest" and material compensation are prohibited—as well as messages to individuals of certain countries who do not permit Amateur Radio message traffic on behalf of others. [97.3a42]

N1I10 If you are allowing a non-amateur friend to use your station to talk to someone in the US, and a foreign station breaks in to talk to your friend, what should you do?
 A. Have your friend wait until you find out if the US has a third-party agreement with the foreign station's government.
 B. Stop all discussions and quickly sign off.
 C. Since you can talk to any foreign amateurs, your friend may keep talking as long as you are the control operator.
 D. Report the incident to the foreign amateur's government.
ANSWER A: We must have a third-party agreement to allow your friend to talk to a foreign amateur radio operator. [97.115a2]

N1I11 When are you allowed to transmit a message to a station in a foreign country for a third party?
 A. Anytime
 B. Never
 C. Anytime, unless there is a third-party agreement between the US and the foreign government

D. If there is a third-party agreement with the US government, or if the third party could be the control operator

ANSWER D: The following list is an example of countries with whom we have a third-party agreement. As you will see, most of our third-party agreements are with South American countries, with a few countries to our west, but almost no countries in Europe. [97.115a2]

Table N1I11. List of Countries Permitting Third-Party Traffic

Country	Call Sign Prefix	Country	Call Sign Prefix	Country	Call Sign Prefix
Antigua and Barbuda	V2	Ecuador	HC	Nicaragua	YN
Argentina	LU	El Salvador	YS	Panama	HP
Australia	VK	The Gambia	C5	Paraguay	ZP
Austria, Vienna	4U1VIC	Ghana	9G	Peru	OA
Belize	V3	Grenada	J3	Philippines	DU
Bolivia	CP	Guatemala	TG	St. Christopher & Nevis	V4
Brazil	PY	Guyana	8R	St. Lucia	J6
Canada	VE, VO, VY	Haiti	HH	St. Vincent & Grenadines	J8
Chile	CE	Honduras	HR	Sierra Leone	9L
Colombia	HK	Israel	4X	Swaziland	3D6
Comoros	D6	Jamaica	6Y	Switzerland, Geneva	4U1ITU
Costa Rica	TI	Jordan	JY	Trinidad and Tobago	9Y
Cuba	CO	Liberia	EL	United Kingdom	GB*
Dominica	J7	Mexico	XE	Uruguay	CX
Dominican Republic	HI	Micronesia	V6	Venezuela	YV

*GB3 excluded

N1J False signals or unidentified communications and malicious interference

N1J01 What is a transmission called that disturbs other communications?

A. Interrupted CW
B. Harmful interference
C. Transponder signals
D. Unidentified transmissions

ANSWER B: Intentionally transmitting over another station already on the air is not courteous, not legal, and out of the spirit of good ham radio operating. [97.3a21]

N1J02 Why is transmitting on a police frequency as a "joke" called harmful interference that deserves a large penalty?

A. It annoys everyone who listens.
B. It blocks police calls which might be an emergency and interrupts police communications.
C. It is in bad taste to communicate with non-amateurs, even as a joke.
D. It is poor amateur practice to transmit outside the amateur bands.

ANSWER B: Many dual-band amateur radio sets are frequency agile after modifications have been done. If you were to transmit on a police frequency, this would be illegal, and it could block a police call for an emergency broadcast. Never play games with a ham radio transceiver. [97.3a21]

N1J03 When may you deliberately interfere with another station's communications?
 A. Only if the station is operating illegally
 B. Only if the station begins transmitting on a frequency you are using
 C. Never
 D. You may expect, and cause, deliberate interference because it can't be helped during crowded band conditions.

ANSWER C: Ham radio operators pride themselves in being polite. Deliberate interference is rare, and will not be tolerated. The FCC will respond to jamming complaints from the amateur community. You could lose your amateur operator/primary station license permanently if found guilty of intentional interference. [97.101d]

N1J04 When may false or deceptive amateur signals or communications be transmitted?
 A. Never
 B. When operating a beacon transmitter in a "fox hunt" exercise
 C. When playing a harmless "practical joke"
 D. When you need to hide the meaning of a message for secrecy

ANSWER A: Going on the air using someone else's call sign is strictly forbidden! [97.113d]

N1J05 If an amateur pretends there is an emergency and transmits the word "MAYDAY," what is this called?
 A. A traditional greeting in May
 B. An emergency test transmission
 C. False or deceptive signals
 D. Nothing special; "MAYDAY" has no meaning in an emergency.

ANSWER C: It's wise not to even utter the word "MAYDAY" in the course of your conversation. Reserve this word for the highest of emergencies over the worldwide bands. On local VHF and UHF repeaters, the equivalent of the worldwide word "MAYDAY" is the phrase "Break, break, break." A triple break signifies a local emergency. [97.113d]

N1J06 When may an amateur transmit unidentified communications?
 A. Only for brief tests not meant as messages
 B. Only if it does not interfere with others
 C. Never, except to control a model craft
 D. Only for two-way or third-party communications

ANSWER C: Use your call sign often on the air. It's required by law, and you should be proud of it. [97.119a]

N1J07 What is an amateur communication called that does not have the required station identification?
 A. Unidentified communications or signals
 B. Reluctance modulation
 C. Test emission
 D. Tactical communication

ANSWER A: Although not required, it's common practice to use your call sign at the beginning of a transmission, too. However, the law does allow you to begin communicating without your call sign for up to 10 minutes. While it's not required at the beginning of a transmission, it's required when you sign off. [97.119a]

N1J08 If you hear a voice distress signal on a frequency outside of your license privileges, what are you allowed to do to help the station in distress?

A. You are NOT allowed to help because the frequency of the signal is outside your privileges.

B. You are allowed to help only if you keep your signals within the nearest frequency band of your privileges.

C. You are allowed to help on a frequency outside your privileges only if you use international Morse code.

D. You are allowed to help on a frequency outside your privileges in any way possible.

ANSWER D: In an emergency, anything goes! If you hear someone calling "MAYDAY" on a frequency outside of your normal operating privileges, it's perfectly okay to transmit on any frequency to save someone's life. [97.405a]

N1J09 If you answer someone on the air without giving your call sign, what type of communication have you just conducted?

A. Test transmission

B. Tactical signal

C. Packet communication

D. Unidentified communication

ANSWER D: Jumping into a conversation with a quick comment still requires proper identification with your FCC call letters. [97.119a]

N1J10 When may you use your amateur station to transmit an "SOS" or "MAYDAY"?

A. Never

B. Only at specific times (at 15 and 30 minutes after the hour)

C. In a life or property threatening emergency

D. When the National Weather Service has announced a severe weather watch

ANSWER C: As mentioned previously, don't even utter the word "MAYDAY" in the course of your conversation. Only in the highest of emergencies would you "MAYDAY" over the worldwide bands. The same applies, of course, for "SOS" when transmitting with CW. [97.403]

N1J11 When may you send a distress signal on any frequency?

A. Never

B. In a life or property threatening emergency

C. Only at specific times (at 15 and 30 minutes after the hour)

D. When the National Weather Service has announced a severe weather watch

ANSWER B: In a life and death situation, you may send out a distress signal on any frequency and any band, regardless of the license you may hold. [97.405a]

Subelement N2 – Operating Procedures

2 exam questions
2 topic groups

N2A Choosing a frequency for tune-up, operating or emergencies; understanding the Morse code; RST signal reports; Q signals; voice communications and phonetics

N2A01 What should you do before you transmit on any frequency?
A. Listen to make sure others are not using the frequency.
B. Listen to make sure that someone will be able to hear you.
C. Check your antenna for resonance at the selected frequency.
D. Make sure the SWR on your antenna feed line is high enough.

ANSWER A: Always listen for a few seconds before initiating a transmitted call. On worldwide, ask, "Is the frequency in use"? Always choose a frequency within your privileges, within the American Radio Relay League (ARRL) suggested band plan, and clear of an ongoing conversation.

N2A02 If you make contact with another station and your signal is extremely strong and perfectly readable, what adjustment might you make to your transmitter?
A. Turn on your speech processor.
B. Reduce your SWR.
C. Continue with your contact, making no changes.
D. Turn down your power output to the minimum necessary.

ANSWER D: Once you "hook up" with another station that is coming in loud and clear, turn down your power to minimize interference to other stations many miles away.

N2A03 What is one way to shorten transmitter tune-up time on the air to cut down on interference?
A. Use a random wire antenna.
B. Tune up on 40 meters first, then switch to the desired band.
C. Tune the transmitter into a dummy load.
D. Use twin lead instead of coaxial-cable feed lines.

ANSWER C: It is sometimes necessary to listen to your own signal with a companion receiver to make adjustments to the transmitter. Instead of putting the signal out on the air, use a dummy load to absorb your signal and to keep it from causing interference during your testing procedures. This may be a simple light bulb, or a 50-ohm non-inductive resistor, capable of handling your power output.

N2A04 If you are in contact with another station and you hear an emergency call for help on your frequency, what should you do?
A. Tell the calling station that the frequency is in use.
B. Direct the calling station to the nearest emergency net frequency.
C. Call your local Civil Preparedness Office and inform them of the emergency.
D. Stop your QSO immediately and take the emergency call.

ANSWER D: An emergency call always has the highest priority. Do what you can to take down the message accurately, and then call the proper authorities.

N2A05 What is the correct way to call CQ when using Morse code?

A. Send the letters "CQ" three times, followed by "DE," followed by your call sign sent once.
B. Send the letters "CQ" three times, followed by "DE," followed by your call sign sent three times.
C. Send the letters "CQ" ten times, followed by "DE," followed by your call sign sent once.
D. Send the letters "CQ" over and over.

ANSWER B: Practice sending "CQ" on a Morse code oscillator before actually going on the air. Don't send it too fast. A response will be at the same rate that you send "CQ."

N2A06 How should you answer a Morse code CQ call?

A. Send your call sign four times.
B. Send the other station's call sign twice, followed by "DE," followed by your call sign twice.
C. Send the other station's call sign once, followed by "DE," followed by your call sign four times.
D. Send your call sign followed by your name, station location and a signal report.

ANSWER B: When you respond to a Morse code "CQ," respond at the rate that the sending station was using. If they were sending fast, then it's okay to send code back to them fast. The letters "QRS" mean "please send more slowly."

N2A07 At what speed should a Morse code CQ call be transmitted?

A. Only speeds below five WPM
B. The highest speed your keyer will operate
C. Any speed at which you can reliably receive
D. The highest speed at which you can control the keyer

ANSWER C: We use CQ for a general call to communicate with anyone, about anything, on worldwide frequencies. When using telegraphy, a fast CW CQ call will result in a very fast reply since it will be assumed that you can receive Morse code as fast as you can send it. Don't transmit CW faster than you can receive it.

N2A08 What is the meaning of the procedural signal "CQ"?

A. "Call on the quarter hour."
B. "New antenna is being tested." (no station should answer)
C. "Only the called station should transmit."
D. "Calling any station."

ANSWER D: CQ is a fun way to establish communications on worldwide bands. We don't use CQ on the VHF or UHF bands. Things are not that formal up there, so use CQ only on the 10-meter, 15-meter, 40-meter, and 80-meter Novice bands.

N2A09 What is the meaning of the procedural signal "DE"?

A. "From" or "this is," as in "W9NGT DE N9BTT"
B. "Directional Emissions" from your antenna
C. "Received all correctly"
D. "Calling any station"

ANSWER A: When working CW, it's much easier to send abbreviations than the whole word or phrase. Abbreviations are very important for you to know for your examinations, both written and code.

N2A10 What is the meaning of the procedural signal "K"?
A. "Any station transmit"
B. "All received correctly"
C. "End of message"
D. "Called station only transmit"

ANSWER A: Telegraphy means using code with a telegraph key. Don't confuse this word with telephony which means using a microphone, like a telephone. In telegraphy, the "K" means the same as "over" or "go ahead." The "SK" indicates the sender is signing off or "goodbye."

N2A11 What is meant by the term "DX"?
A. Best regards
B. Distant station
C. Calling any station
D. Go ahead

ANSWER B: When someone says they are working "DX", this means they are working a distant station that may be many miles away.

N2A12 What is the meaning of the term "73"?
A. Long distance
B. Best regards
C. Love and kisses
D. Go ahead

ANSWER B: The term "73" means best regards, and is an old railroad telegraph signature. Ladies sometimes sign "88". That is an affectionate way of saying their best regards.

N2A13 What are RST signal reports?
A. A short way to describe ionospheric conditions
B. A short way to describe transmitter power
C. A short way to describe signal reception
D. A short way to describe sunspot activity

ANSWER C: The RST signal reporting system is the way we use our worldwide radios to describe signal reception. We normally do not use the RST reporting system on VHF/UHF FM repeater communications. The first digit (R) indicates Readability, the second digit (S) indicates received Signal Strength, and the third digit (T) represents Tone.

N2A14 What does RST mean in a signal report?
A. Recovery, signal strength, tempo
B. Recovery, signal speed, tone
C. Readability, signal speed, tempo
D. Readability, signal strength, tone

ANSWER D: It's up to you and your ear to determine readability on a scale of 1 to 5. Five is best, 3 is fair, and 1 is for the signal that you can just barely make out.

N2A15 What is one meaning of the Q signal "QRS"?
A. Interference from static
B. Send more slowly.
C. Send RST report.
D. Radio station location is

ANSWER B: When you first start on the air, "QRS" can be very important because it means "send more slowly." Remember, in your CW transmissions, it's much easier to send abbreviations; so learn them well, especially for your examinations.

N2A16 What is one meaning of the Q signal "QTH"?
A. Time here is
B. My name is
C. Stop sending
D. My location is

ANSWER D: Example: What is your QTH?

N2A17 What is a QSL card?
A. A letter or postcard from an amateur pen pal
B. A Notice of Violation from the FCC
C. A written proof of communication between two amateurs
D. A postcard reminding you when your license will expire

ANSWER C: Hams exchange colorful postcard QSLs to confirm contacts. Always take a look at another ham's QSL card collection. You will find it fascinating! Here is your author's QSL card.

73–West/Coast Amateur Radio School

WB6NOA

OVS, OBS

Gordon West, Instructor
2414 College Drive
Costa Mesa, CA 92626
Orange County

Life Memb.

144.330 MHz Simplex Anytime

N2A18 What is the correct way to call CQ when using voice?
A. Say "CQ" once, followed by "this is," followed by your call sign spoken three times.
B. Say "CQ" at least five times, followed by "this is," followed by your call sign spoken once.
C. Say "CQ" three times, followed by "this is," followed by your call sign spoken three times.
D. Say "CQ" at least ten times, followed by "this is," followed by your call sign spoken once.

ANSWER C: We use "CQ" as a general call only on the worldwide bands. On the VHF and UHF bands, we are less formal, and simply announce ourselves as being on the air by just giving our call sign.

N2A19 How should you answer a voice CQ call?

A. Say the other station's call sign at least ten times, followed by "this is," then your call sign at least twice.

B. Say the other station's call sign at least five times phonetically, followed by "this is," then your call sign at least once.

C. Say the other station's call sign at least three times, followed by "this is," then your call sign at least five times phonetically.

D. Say the other station's call sign once, followed by "this is," then your call sign given phonetically.

ANSWER D: If you hear a lively CQ call, then go ahead and respond to that station and enjoy a great conversation. If the station is coming in clear, you would only need to call it once, followed by "This is," and then give your call sign, slowly, and phonetically. Your author also likes to give his location.

N2A20 To make your call sign better understood when using voice transmissions, what should you do?

A. Use Standard International Phonetics for each letter of your call.

B. Use any words which start with the same letters as your call sign for each letter of your call.

C. Talk louder.

D. Turn up your microphone gain.

ANSWER A: When you voice your call sign, a "P" might sound like a "B", and that sounds like a "C", or maybe it was "D". To end the confusion, use the phonetic alphabet the first time you make contact with a new station. The standard phonetic alphabet adopted by the International Telecommunication Union is shown in Table N2A20.

Table N2A20. Phonetic Alphabet
Adopted by the International Telecommunication Union

A - Alpha	H - Hotel	O - Oscar	V - Victor
B - Bravo	I - India	P - Papa	W - Whiskey
C - Charlie	J - Juliette	Q - Quebec	X - X-Ray
D - Delta	K - Kilo	R - Romeo	Y - Yankee
E - Echo	L - Lima	S - Sierra	Z - Zulu
F - Foxtrot	M - Mike	T - Tango	
G - Golf	N - November	U - Uniform	

N2B Radio teleprinting; packet; repeater operating procedures; special operations

N2B01 What is the correct way to call CQ when using RTTY?

A. Send the letters "CQ" three times, followed by "DE," followed by your call sign sent once.

B. Send the letters "CQ" three to six times, followed by "DE," followed by your call sign sent three times.

C. Send the letters "CQ" ten times, followed by the procedural signal "DE," followed by your call sent one time.

D. Send the letters "CQ" over and over.

ANSWER B: RadioTeleTYpe, or RTTY, operation involves typewriters actuated by binary-coded radio signals. More home computers are used today by hams to receive RTTY than teleprinters. We send CQ three to six times, rather than three times.

N2B02 What speed should you use when answering a CQ call using RTTY?
- A. Half the speed of the received signal
- B. The same speed as the received signal
- C. Twice the speed of the received signal
- D. Any speed, since RTTY systems adjust to any signal speed

ANSWER B: Operating an RTTY station with your home computer is fun. It's also challenging. Not only do you need to tune in the station correctly, but you also need to match the speed of your equipment with the sending rate of the station on the air. The sending rate of 45 baud is common on worldwide bands.

N2B03 What does "connected" mean in a packet-radio link?
- A. A telephone link is working between two stations.
- B. A message has reached an amateur station for local delivery.
- C. A transmitting station is sending data to only one receiving station; it replies that the data is being received correctly.
- D. A transmitting and receiving station are using a digipeater, so no other contacts can take place until they are finished.

ANSWER C: Your packet radio system gives you many winking lights that indicate your station is interacting with another station.

N2B04 What does "monitoring" mean on a packet-radio frequency?
- A. The FCC is copying all messages.
- B. A member of the Amateur Auxiliary to the FCC's Field Operations Bureau is copying all messages.
- C. A receiving station is displaying all messages sent to it, and replying that the messages are being received correctly.
- D. A receiving station is displaying messages that may not be sent to it, and is not replying to any message.

ANSWER D: Eavesdropping is a favorite practice on packet radio—and no one will know that you are there.

N2B05 What is a digipeater?
- A. A packet-radio station that retransmits only data that is marked to be retransmitted
- B. A packet-radio station that retransmits any data that it receives
- C. A repeater that changes audio signals to digital data
- D. A repeater built using only digital electronics parts

ANSWER A: Long-time ham radio operators have established excellent repeater stations and packet repeater set-ups for everyone to use. We strongly support those hams that give of their time to help you extend your communications range, and we hope that you will support their stations by joining their membership organizations.

N2B06 What does "network" mean in packet radio?
- A. A way of connecting terminal-node controllers by telephone so data can be sent over long distances
- B. A way of connecting packet-radio stations so data can be sent over long distances
- C. The wiring connections on a terminal-node controller board
- D. The programming in a terminal-node controller that rejects other callers if a station is already connected

ANSWER B: Packet stations are hooking up throughout the country into networks covering many states. It may take you a few weeks to figure out the system, but your new enhanced Novice Class license gives you full privileges to take advantage of this new communications medium.

N2B07 What is simplex operation?
A. Transmitting and receiving on the same frequency
B. Transmitting and receiving over a wide area
C. Transmitting on one frequency and receiving on another
D. Transmitting one-way communications

ANSWER A: Simplex means same frequency. Operate simplex on VHF or UHF when the other station is within a few miles of your station. The opposite of simplex is duplex, a type of repeater operation.

N2B08 When should you use simplex operation instead of a repeater?
A. When the most reliable communications are needed
B. When a contact is possible without using a repeater
C. When an emergency telephone call is needed
D. When you are traveling and need some local information

ANSWER B: If you are communicating with a station that is located within 10 miles of you, go "simplex" (or direct) rather than through a repeater. This localizes your transmissions and frees the repeater for more distant contacts.

N2B09 What is a good way to make contact on a repeater?
A. Say the call sign of the station you want to contact three times.
B. Say the other operator's name, then your call sign three times.
C. Say the call sign of the station you want to contact, then your call sign.
D. Say, "Breaker, breaker," then your call sign.

ANSWER C: Before taking advantage of your new repeater privileges, do a lot of listening. This will help you develop good operating technique. Don't use any jargon that you may have used in other radio services. It may take a few weeks to shake your old lingo, but hams have their own new lingo that you will want to learn and use. Simply stating your call sign is a good opener in any situation and on any frequency. If you know who you want to call, say their call sign, then your call sign.

N2B10 When using a repeater to communicate, what do you need to know about the repeater besides its output frequency?
A. Its input frequency
B. Its call sign
C. Its power level
D. Whether or not it has a phone patch

ANSWER A: A repeater is a little bit like a telephone handset—we listen on the output, but we talk on the input. If we tune into a repeater output frequency, we would transmit on its input frequency. Most repeaters use a common offset.

N2B11 What is the main purpose of a repeater?
A. To make local information available 24 hours a day
B. To link amateur stations with the telephone system

C. To retransmit NOAA weather information during severe storm warnings

D. To increase the range of portable and mobile stations

ANSWER D: A repeater (or "machine" as hams commonly call it) retransmits low-powered hand-held, base station or mobile radio signals from tall high-gain antennas at higher power, thereby extending communications range. Repeater communications is the ham's very popular social party line! Novice Class operators have repeater privileges on the 222-MHz and 1270-MHz bands. You should purchase a repeater guide so that you can learn the frequencies of the various repeaters in your area.

N2B12 What does it mean to say that a repeater has an input and an output frequency?

A. The repeater receives on one frequency and transmits on another.

B. The repeater offers a choice of operating frequency, in case one is busy.

C. One frequency is used to control the repeater and another is used to retransmit received signals.

D. The repeater must receive an access code on one frequency before retransmitting received signals.

ANSWER A: Repeater directories publish the repeater frequencies by output. The plus (+) or minus (−) indicates the input "split" that you dial in on your VHF or UHF ham set. A plus (+) indicates a higher input and a minus (−) indicates a lower input. When you start to transmit, your transmitter should automatically go to the proper input frequency. Some repeaters also require a sub-audible tone as part of your input transmission. Ask the local operators how to engage the tone signal on your ham radio set.

N2B13 What is an autopatch?

A. Something that automatically selects the strongest signal to be repeated

B. A device which connects a mobile station to the next repeater if it moves out of range of the first

C. A device that allows repeater users to make telephone calls from their stations

D. A device which locks other stations out of a repeater when there is an important conversation in progress

ANSWER C: Now you have a car phone—and a pocket phone! Many ham repeaters are tied into automatic telephone system interconnections. Once you join a repeater group, you may be given the specific access code for making local phone calls. Remember, it's not legal to call the office or make business phone calls from an Amateur Radio set. Don't confuse your new pleasure telephone patch capabilities with the utility of a regular cellular telephone. For business, you should use cellular telephone or CB. For strictly personal phone calls, use your ham autopatch, and be aware that the whole world is listening! Phone patches are anything but private.

N2B14 What is the purpose of a repeater time-out timer?

A. It lets a repeater have a rest period after heavy use.

B. It logs repeater transmit time to predict when a repeater will fail.

C. It tells how long someone has been using a repeater.

D. It limits the amount of time someone can transmit on a repeater.

ANSWER D: Time-out timers keep repeater operators from getting long-winded. The repeater cycles off the air if it doesn't get a few seconds break during one long transmission. Some timers are as short as 30 seconds. It's always good practice to keep your transmissions shorter than a one-half minute period. If you need to talk longer, announce, "Reset," release the mike button, and let the repeater reestablish its time-out timer.

N2B15 What is a CTCSS (or PL) tone?
A. A special signal used for telecommand control of model craft
B. A sub-audible tone added to a carrier which may cause a receiver to accept a signal
C. A tone used by repeaters to mark the end of a transmission
D. A special signal used for telemetry between amateur space stations and Earth stations

ANSWER B: CTCSS stands for Continuous Tone Coded Squelch System. This is a sub-audible tone that rides along with your FM carrier to access other stations using the same CTCSS tone.

Table N2B15. EIA Standard Subaudible CTCSS (PL) Tone Frequencies

Freq.	Tone No.	Tone Code	Freq.	Tone No.	Tone Code	Freq.	Tone No.	Tone Code
67.0	01	XZ	110.9	15	2Z	179.9	29	6B
71.9	02	XA	114.8	16	2A	186.2	30	7Z
74.4	03	WA	118.8	17	2B	192.8	31	7A
77.0	04	XB	123.0	18	3Z	203.5	32	M1
79.7	05	SP	127.3	19	3A	206.5		8Z
82.5	06	YZ	131.8	20	3B	210.7	33	M2
85.4	07	YA	136.5	21	4Z	218.8	34	M3
88.5	08	YB	141.3	22	4A	225.7	35	M4
91.5	09	ZZ	146.2	23	4B	229.2		9Z
94.8	10	ZA	151.4	24	5Z	233.6	36	
97.4	11	ZB	156.7	25	5A	241.8		M5
100.0	12	1Z	162.2	26	5B	250.3		M6
103.5	13	1A	167.9	27	6Z	256.3		M7
107.2	14	1B	173.8	28	6A			

Courtesy of EIA

Subelement N3 – Radio Wave Propagation
1 exam question
1 topic group

N3A Radio wave propagation, line of sight, ground wave, sky wave, sunspots and the sunspot cycle, reflection of VHF/UHF signals

N3A01 When a signal travels in a straight line from one antenna to another, what is this called?
A. Line-of-sight propagation
B. Straight-line propagation
C. Knife-edge diffraction
D. Tunnel propagation

ANSWER A: Signals on your 222-MHz and 1270-MHz bands travel in straight lines. Atmospheric refraction may cause the signals to extend 20 percent beyond the optical horizon.

N3A02 What type of propagation usually occurs from one hand-held VHF transceiver to another nearby?

A. Tunnel propagation
B. Sky-wave propagation
C. Line-of-sight propagation
D. Auroral propagation

ANSWER C: These line-of-sight signals bounce off of buildings, and many times will even transmit into elevators. While signals travel in straight lines, they often take some great bounces. They also are bent by local still-air atmospheric conditions.

N3A03 How do VHF and UHF radio waves usually travel from a transmitting antenna to a receiving antenna?

A. They bend through the ionosphere.
B. They go in a straight line.
C. They wander in any direction.
D. They move in a circle going either east or west from the transmitter.

ANSWER B: VHF and UHF radio signals travel line of sight. They do not bend through the ionosphere.

N3A04 What can happen to VHF or UHF signals going towards a metal-framed building?

A. They will go around the building.
B. They can be bent by the ionosphere.
C. They can be easily reflected by the building.
D. They are sometimes scattered in the ectosphere.

ANSWER C: Your small 222-MHz or 1270-MHz hand-held transceiver will work nicely inside buildings and even in elevators. Signals are so reflective at these frequencies that they manage to find a way out to that distant repeater. You may also find that there are different areas in your house or office that are better than others for working distant repeaters because of nearby reflections.

N3A05 When a signal travels along the surface of the Earth, what is this called?

A. Sky-wave propagation
B. Knife-edge diffraction
C. E-region propagation
D. Ground-wave propagation

ANSWER D: All stations on all frequencies emit ground waves that hug the earth. They travel out from your transmitter antenna up to approximately 100 miles. See figure at question N3A06.

N3A06 How does the range of sky-wave propagation compare to ground-wave propagation?

A. It is much shorter.
B. It is much longer.
C. It is about the same.
D. It depends on the weather.

ANSWER B: Ground waves have a much shorter range than sky waves reflected off the ionosphere. It's quite common for sky-wave signals on 10 meters originating 3000 miles away to overpower local signals originating only a few miles away! You will be fascinated with 10-meter propagation.

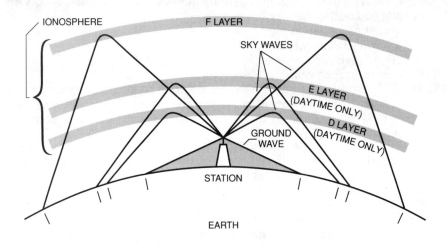

Radio Wave Propagation

Source: *Mobile 2-Way Radio Communications,* G.West, © 1993, Master Publishing, Inc.

N3A07 When a signal is returned to earth by the ionosphere, what is this called?

A. Sky-wave propagation
B. Earth-moon-earth propagation
C. Ground-wave propagation
D. Tropospheric propagation

ANSWER A: What you couldn't do on CB (Citizen's Band) is a way of life on ham radio! Radio waves are bounced off of the ionospheric E or F layer and may return to earth thousands of miles away! This 10-meter propagation occurs in daylight and evening hours—seldom beyond 9:00 PM. The ionosphere becomes mirror-like due to ultraviolet radiation given off by the sun. The mirror may be so intense during daylight hours that 10-meter signals may travel halfway around the world. DXing will be fun for you as a Novice!

N3A08 What is the usual cause of sky-wave propagation?

A. Signals are reflected by a mountain.
B. Signals are reflected by the moon.
C. Signals are bent back to earth by the ionosphere.
D. Signals are repeated by a repeater.

ANSWER C: On the 10-meter band, sky waves will give you some phenomenal long-range contacts during daylight and twilight hours. Sorry folks, no sky waves on the 222-MHz or 1270-MHz band.

N3A09 What is a skip zone?

A. An area covered by ground-wave propagation
B. An area covered by sky-wave propagation
C. An area which is too far away for ground-wave propagation, but too close for sky-wave propagation
D. An area which is too far away for ground-wave or sky-wave propagation

ANSWER C: Increasing your power won't help establish communications with a station that is being missed within your skip zone. If you switch to other bands, chances are good that lower frequencies (longer wavelength) may skip in closer.

N3A10 What are the regions of ionized gases high above the earth called?
A. The ionosphere
B. The troposphere
C. The gas region
D. The ion zone

ANSWER A: The region of ionized gases high above the earth is called the ionosphere.

N3A11 How do sunspots change the ionization of the atmosphere?
A. The more sunspots there are, the greater the ionization
B. The more sunspots there are, the less the ionization
C. Unless there are sunspots, the ionization is zero
D. They have no effect

ANSWER A: The higher the solar activity, the greater the ionization. Listen to WWV, 10 or 15 MHz, at 18 minutes past every hour. They will report on the amount of solar activity, and you can many times predict whether 10 meters is going to be hot, or not, the next day.

N3A12 How long is an average sunspot cycle?
A. 2 years
B. 5 years
C. 11 years
D. 17 years

ANSWER C: The sun exhibits periods of high solar activity, peaking every 11 years. During the peak of the sunspot cycles, you will be able to reach out with your 10-meter voice privileges to the most distant stations worldwide. At the minimum of the next sunspot cycle (around 1997), solar activity will drop off, and 10-meter long range contacts may only be a summertime phenomena.

Subelement N4 – Amateur Radio Practices

4 exam questions
4 topic groups

N4A Unauthorized use prevention, lightning protection, and station grounding

N4A01 How could you best keep unauthorized persons from using your amateur station at home?
A. Use a carrier-operated relay in the main power line.
B. Use a key-operated on/off switch in the main power line.
C. Put a "Danger - High Voltage" sign in the station.
D. Put fuses in the main power line.

ANSWER B: You are responsible for unauthorized use of your station even though the use may be without your approval. Most hams remove the microphone and shut off the main power line to protect a station from being operated without permission. (Before you can go on the air again, you must remember where you hid the mike.)

N4A02 How could you best keep unauthorized persons from using a mobile amateur station in your car?
A. Disconnect the microphone when you are not using it.
B. Put a "do not touch" sign on the radio.
C. Turn the radio off when you are not using it.
D. Tune the radio to an unused frequency when you are done using it.
ANSWER A: One good way to keep unauthorized persons from using a mobile amateur station is to disconnect the microphone. Lock it up in the glove compartment.

N4A03 Why would you use a key-operated on/off switch in the main power line of your station?
A. To keep unauthorized persons from using your station
B. For safety, in case the main fuses fail
C. To keep the power company from turning off your electricity during an emergency
D. For safety, to turn off the station in the event of an emergency
ANSWER A: It is a good idea to consider some way of securely turning off your set with a secret switch so no one can accidentally or purposefully energize it and use it.

N4A04 Why should you ground all antenna and rotator cables when your amateur station is not in use?
A. To lock the antenna system in one position
B. To avoid radio frequency interference
C. To save electricity
D. To protect the station and building from lightning damage
ANSWER D: An easy way to provide an automatic antenna grounding circuit is to use a multi-position antenna selector switch. Purposely ground one of the connections so that you can switch to it when your equipment is not in use.

N4A05 How can an antenna system best be protected from lightning damage?
A. Install a balun at the antenna feed point.
B. Install an RF choke in the antenna feed line.
C. Ground all antennas when they are not in use.
D. Install a fuse in the antenna feed line.
ANSWER C: Always use a good quality amateur-grade grounded antenna selector switch for lightning protection. You also can install lightning arrestors.

N4A06 How can amateur station equipment best be protected from lightning damage?
A. Use heavy insulation on the wiring.
B. Never turn off the equipment.
C. Disconnect the ground system from all radios.
D. Disconnect all equipment from the power lines and antenna cables.
ANSWER D: If you suspect a lightning storm approaching, unplug everything, and actually remove the set from other grounded equipment in your ham shack.

N4A07 For best protection from electrical shock, what should be grounded in an amateur station?
 A. The power supply primary
 B. All station equipment
 C. The antenna feed line
 D. The AC power mains
ANSWER B: When you add new gear to your ham station setup, always connect the chassis to the other gear with a good common braid or foil connection. Be certain all connections are tight! You'll have less ground noise that can affect transmitted and received signals.

N4A08 What is usually a good indoor grounding point for an amateur station?
 A. A metallic cold water pipe
 B. A plastic cold water pipe
 C. A window screen
 D. A metallic natural gas pipe
ANSWER A: Run your copper foil from the chassis of each piece of equipment to a common grounding bar (a length of 2-inch diameter copper water pipe is ideal). Connect the grounding bar to a good earth ground, either a bar driven into the ground for at least eight feet, or a cold water pipe that goes underground. Avoid hot water pipes since they do not run directly to earth. Don't use a cold water pipe grounding system if you have PVC plastic water pipes! Instead, use a 6- to 8-foot copper-plated steel ground rod that's available from Radio Shack.

N4A09 Where should you connect the chassis of each piece of your station equipment to best protect against electrical shock?
 A. To insulated shock mounts
 B. To the antenna
 C. To a good ground connection
 D. To a circuit breaker
ANSWER C: A good earth ground connection will minimize the potential of an electrical shock if it is well connected to the chassis of each piece of station equipment. We normally use ground foil for grounding purposes.

N4A10 Which of these materials is best for a ground rod driven into the earth?
 A. Hard plastic
 B. Copper or copper-clad steel
 C. Iron or steel
 D. Fiberglass
ANSWER B: Be careful where you drive in your copper/clad steel ground rod. One time your author managed to pound it through a gas pipe. What a big surprise!

N4A11 If you ground your station equipment to a ground rod driven into the earth, what is the shortest length the rod should be?
 A. 4 feet
 B. 6 feet
 C. 8 feet
 D. 10 feet

ANSWER C: Use a sledge hammer to drive the ground rod at least eight feet into the earth. Always wear protective goggles when hammering away at the rod. Gloves are a good idea as well. They will help to prevent blisters.

N4B Radio frequency safety precautions, safety interlocks, antenna installation safety procedures

N4B01 What should you do for safety when operating at 1270 MHz?
    A. Make sure that an RF leakage filter is installed at the antenna feed point.
    B. Keep antenna away from your eyes when RF is applied.
    C. Make sure the standing wave ratio is low before you conduct a test.
    D. Never use a shielded horizontally polarized antenna.

ANSWER B: A waveguide is a form of low-loss transmission line that transports microwave energy from one place to another much like a garden hose carries water. The RF radiation bounces off of the sides of the waveguide along its route. Concentrated microwave radiation is generally considered to be biologically harmful to humans—particularly to the eyes, so don't needlessly expose yourself to this hazard. Be sure to keep your face away from the antenna or waveguide at these frequencies.

N4B02 What should you do for safety if you put up a UHF transmitting antenna?
    A. Make sure the antenna will be in a place where no one can get near it when you are transmitting.
    B. Make sure that RF field screens are in place.
    C. Make sure the antenna is near the ground to keep its RF energy pointing in the correct direction.
    D. Make sure you connect an RF leakage filter at the antenna feed point.

ANSWER A: Antennas are dangerous when you are transmitting. Even though you may be putting out just a few watts of power, ultra high frequency (UHF) radio waves may be harmful to your body within a few inches of the antenna system. Keep everyone well clear.

N4B03 What should you do for safety before removing the shielding on a UHF power amplifier?
    A. Make sure all RF screens are in place at the antenna feed line.
    B. Make sure the antenna feed line is properly grounded.
    C. Make sure the amplifier cannot accidentally be turned on.
    D. Make sure that RF leakage filters are connected.

ANSWER C: If you are working on anything with high voltage, or potential amounts of UHF output, be cautious. Make absolutely sure no one can accidentally turn on the equipment while your fingers are inside.

N4B04 Why should you use only good quality coaxial cable and connectors for a UHF antenna system?
    A. To keep RF loss low
    B. To keep television interference high

C. To keep the power going to your antenna system from getting too high

D. To keep the standing wave ratio of your antenna system high

ANSWER A: When assembling your VHF and UHF antenna system, use rigid coaxial cable or the highest quality and best coax you can purchase at the ham radio store. Never use coaxial cable intended for use at lower frequencies.

N4B05 Why should you make sure the antenna of a hand-held transceiver is not close to your head when transmitting?

A. To help the antenna radiate energy equally in all directions

B. To reduce your exposure to the radio-frequency energy

C. To use your body to reflect the signal in one direction

D. To keep static charges from building up

ANSWER B: Since they have attached antennas, hand-held VHF/UHF ham transceivers expose users to the most radio frequency energy. Aim the antenna away from your head and eyes as much as possible!

N4B06 Microwave oven radiation is similar to what type of amateur station RF radiation?

A. Signals in the 3.5 MHz range

B. Signals in the 21 MHz range

C. Signals in the 50 MHz range

D. Signals in the 1270 MHz range

ANSWER D: Way up at the 1270-MHz band, where Novice Class power is limited to 5 watts, you will find microwave radiation similar to microwave oven radiation. Steer clear of this!

N4B07 Why would there be a switch in a high-voltage power supply to turn off the power if its cabinet is opened?

A. To keep dangerous RF radiation from leaking out through an open cabinet

B. To keep dangerous RF radiation from coming in through an open cabinet

C. To turn the power supply off when it is not being used

D. To keep anyone opening the cabinet from getting shocked by dangerous high voltages

ANSWER D: Most linear amplifiers incorporate an automatic disconnect switch when you lift the cover. This usually shorts out the high voltage that may still remain in big capacitors. Never circumvent this switch—it's there for your safety.

N4B08 What kind of safety equipment should you wear if you are working on an antenna tower?

A. A grounding chain

B. A reflective vest of approved color

C. A flashing red, yellow or white light

D. A carefully inspected safety belt, hard hat and safety glasses

ANSWER D: Not all safety belts are safe. The old leather lineman's belt is probably brittle and may break when you are up on the tower. Always check out your climbing equipment well before going aloft.

N4B09 Why should you wear a safety belt if you are working on an antenna tower?
- A. To safely hold your tools so they don't fall and injure someone on the ground
- B. To keep the tower from becoming unbalanced while you are working
- C. To safely bring any tools you might use up and down the tower
- D. To prevent you from accidentally falling

ANSWER D: Your author has slipped on a rung of a tower, and the safety belt saved his life. It could save your life, too.

N4B10 For safety, how high should you place a horizontal wire antenna?
- A. High enough so that no one can touch any part of it from the ground
- B. As close to the ground as possible
- C. Just high enough so you can easily reach it for adjustments or repairs
- D. Above high-voltage electrical lines

ANSWER A: You don't need much height for a simple 10-meter dipole. However, keep it at least 15 feet off the ground. This gives you better range, and it also prevents anyone from accidentally touching it.

N4B11 Why should you wear a hard hat if you are on the ground helping someone work on an antenna tower?
- A. So you won't be hurt if the tower should accidentally fall
- B. To keep RF energy away from your head during antenna testing
- C. To protect your head from something dropped from the tower
- D. So someone passing by will know that work is being done on the tower and will stay away

ANSWER C: Local hams may invite you to an "antenna party." As a group, you may help another ham put up the antenna atop a tower. If you are on the ground, make sure you wear a hard hat and safety glasses.

N4C SWR meaning and measurements

N4C01 What instrument is used to measure standing wave ratio?
- A. An ohmmeter
- B. An ammeter
- C. An SWR meter
- D. A current bridge

ANSWER C: Standing wave ratio is the comparison of the power going forward from the transmitter to the power reflected back from the load—usually an antenna. It is measured by a standing wave ratio bridge—sometimes called a reflectometer, since it measures reflected power. SWR bridges for use below 30 MHz can be purchased for as little as $30.00. For VHF/UHF frequencies, a more expensive (about $150.00) bridge must be used for maximum accuracy.

N4C02 What instrument is used to measure the relative impedance match between an antenna and its feed line?
- A. An ammeter
- B. An ohmmeter
- C. A voltmeter
- D. An SWR meter

ANSWER D: Impedance is similar to the back pressure that a tuned muffler presents to an engine in a sports car. We measure the impedance match between the "back pressure" of your transmitter and antenna with a standing wave ratio (SWR) meter.

N4C03 Where would you connect an SWR meter to measure standing wave ratio?
    A. Between the feed line and the antenna
    B. Between the transmitter and the power supply
    C. Between the transmitter and the receiver
    D. Between the transmitter and the ground

ANSWER A: The very best place to measure the standing wave ratio of your antenna system is between your coax and the antenna connection. This means you need to go up on the roof for the very best measurement. They now have portable SWR analyzers that allow you to do this without a radio transmissions necessary at the other end of the coax.

N4C04 What does an SWR reading of 1:1 mean?
    A. An antenna for another frequency band is probably connected.
    B. The best impedance match has been attained.
    C. No power is going to the antenna.
    D. The SWR meter is broken.

ANSWER B: A 1:1 SWR is a perfect match. However, it doesn't necessarily mean you're going to have a great signal on the airwaves. You can achieve a 1:1 SWR by transmitting into a non-radiating dummy load!

N4C05 What does an SWR reading of less than 1.5:1 mean?
    A. An impedance match which is too low
    B. An impedance mismatch; something may be wrong with the antenna system
    C. A fairly good impedance match
    D. An antenna gain of 1.5

ANSWER C: If you have a 1.5:1 SWR, not bad! But again, this reading is meaningless as a guarantee of good range. Location, height, and surroundings all contribute to the effectiveness of the antenna once you make sure you are transferring maximum power from your transmitter to the antenna by a good impedance match.

N4C06 What does an SWR reading of 4:1 mean?
    A. An impedance match which is too low
    B. An impedance match which is good, but not the best
    C. An antenna gain of 4
    D. An impedance mismatch; something may be wrong with the antenna system

ANSWER D: A 4:1 reading is a sure indication that something is wrong with your antenna system. Here the SWR surely indicates you won't get many contacts. Find out what's wrong and correct it.

N4C07 What kind of SWR reading may mean poor electrical contact between parts of an antenna system?
    A. A jumpy reading
    B. A very low reading

C. No reading at all

D. A negative reading

ANSWER A: It's a good idea to watch your SWR meter while transmitting—especially if the wind is strong. If the wind is blowing your antenna around, you may notice erratic or intermittent readings on your SWR bridge. Guess what? You have a bad connection aloft.

N4C08 What does a very high SWR reading mean?

A. The antenna is the wrong length, or there may be an open or shorted connection somewhere in the feed line.

B. The signals coming from the antenna are unusually strong, which means very good radio conditions.

C. The transmitter is putting out more power than normal, showing that it is about to go bad.

D. There is a large amount of solar radiation, which means very poor radio conditions.

ANSWER A: Another cause of high SWR, not mentioned in the answer, is your antenna is too close to the earth. The 10-meter dipole for your Novice Class voice privileges should be at least 15 feet above the ground. It won't work on the ground.

N4C09 If an SWR reading at the low frequency end of an amateur band is 2.5:1, and is 5:1 at the high frequency end of the same band, what does this tell you about your 1/2-wavelength dipole antenna?

A. The antenna is broadbanded.

B. The antenna is too long for operation on the band.

C. The antenna is too short for operation on the band.

D. The antenna is just right for operation on the band.

ANSWER B: If you go slightly higher in frequency, and the SWR continues to climb, your antenna is too long. Cut off an inch and see what happens.

N4C10 If an SWR reading at the low frequency end of an amateur band is 5:1, and 2.5:1 at the high frequency end of the same band, what does this tell you about your 1/2-wavelength dipole antenna?

A. The antenna is broadbanded.

B. The antenna is too long for operation on the band.

C. The antenna is too short for operation on the band.

D. The antenna is just right for operation on the band.

ANSWER C: Here is just the opposite situation from question N4C09—as you tune higher in frequency within your band limits, the SWR gets better and better. Lengthen the antenna slightly for a better match. If you want to continue to operate at the low end of the band, lengthen the antenna more. Remember "lower-longer" as you tune to get better SWR at the low end of the band.

N4C11 If you use a 3-30 MHz RF-power meter at UHF frequencies, how accurate will its readings be?

A. They may not be accurate at all.

B. They will be accurate enough to get by.

C. They will be accurate but the readings must be divided by two.

D. They will be accurate but the readings must be multiplied by two.

ANSWER A: Any type of RF test equipment for 3-30 MHz operation will normally not be accurate at VHF and UHF frequencies. This means you can't use a CB-type power meter on the VHF and UHF bands.

N4D RFI and its complications

N4D01 What is meant by receiver overload?
A. Too much voltage from the power supply
B. Too much current from the power supply
C. Interference caused by strong signals from a nearby transmitter
D. Interference caused by turning the volume up too high

ANSWER C: Many times a high-pass filter installed at the antenna input of a television will help reduce television interference (TVI) caused by a strong, nearby signal. It is the responsibility of the set owner—not the ham operator—to install this filter.

N4D02 What is one way to tell if radio-frequency interference to a receiver is caused by front-end overload?
A. If connecting a low-pass filter to the transmitter greatly cuts down the interference
B. If the interference is about the same no matter what frequency is used for the transmitter
C. If connecting a low-pass filter to the receiver greatly cuts down the interference
D. If grounding the receiver makes the problem worse

ANSWER B: If you mount your transmitting antenna too close to other receiving antennas used for home entertainment equipment, chances are your transmitted signal will come in on your radio or television receiver. If the interference is independent of your transmitter's location, or where the receiver or TV is tuned, the problem is proximity of the two antennas. This is called "front-end overload." If you see a lot of outside television antennas on roofs around you, chances are your next door neighbors will need to use a high-pass filter on their TV set to prevent interference from your signal. High-pass filters reject all frequencies below a cutoff frequency. The cutoff point (usually around 45 MHz) is higher than your 10-meter transmitted ham signals, but lower than the 55.25 MHz (channel 2) of the first VHF television channel.

N4D03 If your neighbor reports television interference whenever you are transmitting from your amateur station, no matter what frequency band you use, what is probably the cause of the interference?
A. Too little transmitter harmonic suppression
B. Receiver VR tube discharge
C. Receiver overload
D. Incorrect antenna length

ANSWER C: Receiver overload may be reduced by keeping your transmitting antenna as far away from other antennas as possible. Reducing power will also help. Good grounding techniques will help. Staying off the air during big ball games may also help!

N4D04 If your neighbor reports television interference on one or two channels only when you are transmitting on the 15-meter band, what is probably the cause of the interference?

A. Too much low-pass filtering on the transmitter
B. De-ionization of the ionosphere near your neighbor's TV antenna
C. TV receiver front-end overload
D. Harmonic radiation from your transmitter

ANSWER D: If your neighbor is using an outside TV antenna, they might experience multiples of your transmitted signal on the 15-meter and 10-meter bands. These multiples, called harmonics, are minimized by a low-pass filter on your worldwide ham set. Another cure is to encourage your neighbors to sign up for cable television which is less susceptible to interference.

N4D05 What type of filter should be connected to a TV receiver as the first step in trying to prevent RF overload from an amateur HF station transmission?
A. Low-pass
B. High-pass
C. Band pass
D. Notch

ANSWER B: The TV band is higher than your worldwide radio and your new voice privileges on 10 meters. Putting a high-pass filter on the TV receiver, if it's connected to an outside antenna, is a good start in cleaning up interference. Don't put a high-pass filter into a cable TV system. Cable television feeds will not work through a high-pass filter.

N4D06 What type of filter might be connected to an amateur HF transmitter to cut down on harmonic radiation?
A. A key-click filter
B. A low-pass filter
C. A high-pass filter
D. A CW filter

ANSWER B: Low-pass filters are designed only for transceivers used over the bands from 80 meters to 10 meters (the worldwide sets used on 10 meters). Never put a low-pass filter on your 222-MHz or 1270-MHz sets because these bands are much higher than low-pass frequencies. Your VHF and UHF FM equipment seldom interferes with TVs or hi-fi stereo sets.

N4D07 What is meant by harmonic radiation?
A. Unwanted signals at frequencies which are multiples of the fundamental (chosen) frequency
B. Unwanted signals that are combined with a 60-Hz hum
C. Unwanted signals caused by sympathetic vibrations from a nearby transmitter
D. Signals which cause skip propagation to occur

ANSWER A: Every transmitted signal contains a weak second harmonic. If you operate at 7.1 MHz, your second harmonic will be 14.2 MHz, which is in another ham band. If you operate at 28.2 MHz, your second harmonic will land at 56.4 MHz, which is in the middle of television channel 2. Be aware of where your second harmonic falls to be a good operator and a good neighbor!

N4D08 Why is harmonic radiation from an amateur station not wanted?
A. It may cause interference to other stations and may result in out-of-band signals.
B. It uses large amounts of electric power.

    C. It may cause sympathetic vibrations in nearby transmitters.

    D. It may cause auroras in the air.

ANSWER A: A modern ham radio equipment for worldwide operation will contain harmonic suppressors, and so do power amplifiers. Sometimes an external low-pass filter for your high-frequency radio will reduce the second harmonic. Low-pass filters have little insertion signal loss up to a certain cutoff frequency above your 10-meter transmitted signal. They will reject the higher harmonic or unwanted extraneous frequencies. One of the best ways to reduce harmonic interference, however, is to simply reduce your transmitter power level. Don't run more power than you need. There is no point in running a linear amplifier if a "barefoot" 100 watter does just as well—and most will! You will be surprised how little signal gain there is between 100 and 1000 watts! There could be a very big difference in TVI levels, however.

N4D09 What type of interference may come from a multi-band antenna connected to a poorly tuned transmitter?

    A. Harmonic radiation

    B. Auroral distortion

    C. Parasitic excitation

    D. Intermodulation

ANSWER A: Today's modern transistorized transceivers do not require power-amplifier tuning; however, older rigs that use tubes in the final stage require tuning. Improperly tuned transmitters which feed into multi-band antennas can result in harmonic radiation. Tube rigs are fine, but the new solid-state transistorized sets are best, and you won't have to worry about proper tune-up procedures.

N4D10 What is the main purpose of shielding in a transmitter?

    A. It gives the low-pass filter a solid support.

    B. It helps the sound quality of transmitters.

    C. It prevents unwanted RF radiation.

    D. It helps keep electronic parts warmer and more stable.

ANSWER C: Never operate your equipment with the metal cabinet removed. If you are working on your equipment, always reassemble it completely before going on the air. Transmitting with the covers off could lead to unwanted RF radiation escaping from circuits that should be enclosed.

N4D11 If you are told that your amateur station is causing television interference, what should you do?

    A. First make sure that your station is operating properly, and that it does not cause interference to your own television.

    B. Immediately turn off your transmitter and contact the nearest FCC office for assistance.

    C. Connect a high-pass filter to the transmitter output and a low-pass filter to the antenna-input terminals of the television.

    D. Continue operating normally, because you have no reason to worry about the interference.

ANSWER A: On 10 meters with your new Novice Class voice privileges, chances are you may generate some television interference. This is minimized by low-pass filters on your ham set, high-pass filters on your neighbor's TV antenna and receiver, or a good cable company feed system. Start out with your own television—see if you are affecting it.

Subelement N5 – Electrical Principles

N5A Metric prefixes, i.e. pico, micro, milli, centi, kilo, mega, giga

N5A01 If a dial marked in kilohertz shows a reading of 7125 kHz, what would it show if it were marked in megahertz?
    A. 0.007125 MHz
    B. 7.125 MHz
    C. 71.25 MHz
    D. 7,125,000 MHz
ANSWER B: One megahertz (1×10^6) is equal to 1000 kilohertz ($1000 \times 1 \times 10^3$). To convert kHz to MHz, move the decimal point 3 places to the left—7125 kHz is 7.125 MHz. If you want to convert MHz to meters; (i.e., find the wavelength of a frequency), divide MHz into 300.

N5A02 If a dial marked in megahertz shows a reading of 3.525 MHz, what would it show if it were marked in kilohertz?
    A. 0.003525 kHz
    B. 35.25 kHz
    C. 3525 kHz
    D. 3,525,000 kHz
ANSWER C: Move the decimal point 3 places to the right to convert MHz to kHz. 3.525 MHz is 3525 kHz.

N5A03 If a dial marked in kilohertz shows a reading of 3725 kHz, what would it show if it were marked in hertz?
    A. 3,725 Hz
    B. 37.25 Hz
    C. 3,725 Hz
    D. 3,725,000 Hz
ANSWER D: Here's another silly answer you will probably never encounter out there in the ham radio world. Kilo stands for 1000 (1×10^3). Thus, 3725 kHz is really $3725 \times 1 \times 10^3$ or 3,725,000 hertz. Hertz (Hz) means cycles per second.

N5A04 How long is an antenna that is 400 centimeters long?
    A. 0.0004 meters
    B. 4 meters
    C. 40 meters
    D. 40,000 meters
ANSWER B: There are 100 centimeters in one meter, so divide centimeters by 100 to convert to meters. Or move the decimal point 2 places to the left. Calculator keystrokes are: CLEAR 400 ÷ 100 = and the answer is 4.

N5A05 If an ammeter marked in amperes is used to measure a 3000-milliampere current, what reading would it show?
    A. 0.003 amperes
    B. 0.3 amperes
    C. 3 amperes
    D. 3,000,000 amperes

ANSWER C: One milliampere equals one one-thousandth of an ampere (1 × 10^{-3}); therefore, one ampere equal 1000 milliamperes. Divide milliamperes by 1000 to convert to amperes. Or move the decimal point 3 places to the left. Calculator keystrokes are: CLEAR 3000 ÷ 1000 = and the answer is 3.

N5A06 If a voltmeter marked in volts is used to measure a 3500-millivolt potential, what reading would it show?

   A. 0.35 volts
   B. 3.5 volts
   C. 35 volts
   D. 350 volts

ANSWER B: Since milli means 1/1000, one volt equals 1000 millivolts. Move the decimal point 3 places to the left to convert millivolts to volts. This is the same as dividing millivolts by 1000. Calculator keystrokes are: CLEAR 3500 ÷ 1000 = and the answer is 3.5.

N5A07 How many farads is 500,000 microfarads?

   A. 0.0005 farads
   B. 0.5 farads
   C. 500 farads
   D. 500,000,000 farads

ANSWER B: A microfarad is one millionth (1 × 10^{-6}) of a farad. One farad equals one million microfarads. Each place that the decimal point is moved to the left is equivalent to dividing by 10. Move the decimal point 6 places to the left to convert from microfarads to farads. This is the same as dividing microfarads by 1,000,000. Calculator keystrokes are: CLEAR 500000 ÷ 1000000 = and the answer is 0.5.

N5A08 How many microfarads is 1,000,000 picofarads?

   A. 0.001 microfarads
   B. 1 microfarad
   C. 1,000 microfarads
   D. 1,000,000,000 microfarads

ANSWER B: A picofarad is one millionth (1 × 10^{-6}) of a microfarad or one million millionth (1 × 10^{-6} × 1 × 10^{-6} = 1 × 10^{-12}) of a farad. Move the decimal point 6 places to the left to convert to microfarads or 12 places to the left to convert to farads. 1,000,000 picofarads = 1 × 10^{6} × 1 × 10^{-12} = 1 × 10^{-6} of a farad, or 1 microfarad, and 1 microfarad = 0.000001 farad.

N5A09 How many hertz are in a kilohertz?

   A. 10
   B. 100
   C. 1000
   D. 1000000

ANSWER C: Kilo means one thousand (1 × 10^{3}). A kilohertz is 1000 Hz.

N5A10 How many kilohertz are in a megahertz?

   A. 10
   B. 100
   C. 1000
   D. 1000000

ANSWER C: 1 MHz is a thousand kilohertz. Mega means one million (1 × 10^{6}).

N5A11 If you have a hand-held transceiver which puts out 500 milliwatts, how many watts would this be?
- A. 0.02
- B. 0.5
- C. 5
- D. 50

ANSWER B: Remember milli means 1×10^{-3}, or one thousandth, so divide 500 by 1000 to get 0.5 watt. Most hand-helds, on lower power, are rated in milliwatts. A unit rated at 500 mW puts out 0.5 watt. One-half watt is the same as 0.5 watt, and this is plenty for a hand-held.

N5B Concepts of current, voltage, conductor, insulator, resistance, and the measurements thereof

N5B01 What is the flow of electrons in an electric circuit called?
- A. Voltage
- B. Resistance
- C. Capacitance
- D. Current

ANSWER D: Think of the flow of electrons as the flow of water in a stream. If you get out there in midstream, you will feel the current.

N5B02 What is the basic unit of electric current?
- A. The volt
- B. The watt
- C. The ampere
- D. The ohm

ANSWER C: The flow of electrons in a conductor is called current. Current is measured in amperes. Amperes is often referred to as "amps."

N5B03 What is the pressure that forces electrons to flow through a circuit?
- A. Magnetomotive force, or inductance
- B. Electromotive force, or voltage
- C. Farad force, or capacitance
- D. Thermal force, or heat

ANSWER B: Electronic circuits contain electromotive force, or voltage. This causes electrons to flow.

N5B04 What is the basic unit of voltage?
- A. The volt
- B. The watt
- C. The ampere
- D. The ohm

ANSWER A: Volts are volts, the same as voltage!

N5B05 How much voltage does an automobile battery usually supply?
- A. About 12 volts
- B. About 30 volts
- C. About 120 volts
- D. About 240 volts

ANSWER A: Most automobile batteries supply 12 volts direct current.

N5B06 How much voltage does a wall outlet usually supply (in the US)?
A. About 12 volts
B. About 30 volts
C. About 120 volts
D. About 480 volts

ANSWER C: Most voltage coming out of a wall outlet in the United States is about 120 volts, 60 Hz, commonly referred to as 120 VAC. The "AC" means "alternating current."

N5B07 What are three good electrical conductors?
A. Copper, gold, mica
B. Gold, silver, wood
C. Gold, silver, aluminum
D. Copper, aluminum, paper

ANSWER C: Most wire is copper, and this is a good conductor. Some relays use gold- or silver-plated contacts, and these are also good conductors. You can use aluminum foil as a ground plane; it also is a good conductor. Always read all answers completely—mica, wood and paper are insulators!

N5B08 What are four good electrical insulators?
A. Glass, air, plastic, porcelain
B. Glass, wood, copper, porcelain
C. Paper, glass, air, aluminum
D. Plastic, rubber, wood, carbon

ANSWER A: In order for there to be current from one point to another in a circuit, current must have a good completed path of conductivity. If there is a poor or open connection, there will be little or no current.

N5B09 What does an electrical insulator do?
A. It lets electricity flow through it in one direction.
B. It does not let electricity flow through it.
C. It lets electricity flow through it when light shines on it.
D. It lets electricity flow through it.

ANSWER B: An insulator has infinite high resistance so electricity cannot flow through it. It looks like an open circuit.

N5B10 What limits the amount of current that flows through a circuit if the voltage stays the same?
A. Reliance
B. Reactance
C. Saturation
D. Resistance

ANSWER D: In a river, there is a limit as to how much current will flow downstream. Logs and rocks offer "resistance" to the river current. Similarly, certain materials or small diameter wires in an electric circuit offer resistance to electric current.

N5B11 What is the basic unit of resistance?
A. The volt
B. The watt
C. The ampere
D. The ohm

ANSWER D: The basic unit of resistance is called the ohm. If one volt dc is applied to a circuit and one ampere of current results, the circuit has one ohm of resistance.

N5C Ohm's Law (any calculations will be kept to a very low level — no fractions or decimals) and the concepts of energy and power, and open and short circuits

---

N5C01 What formula shows how voltage, current and resistance relate to each other in an electric circuit?
- A. Ohm's Law
- B. Kirchhoff's Law
- C. Ampere's Law
- D. Tesla's Law

ANSWER A: The relationship between voltage, current, and resistance in an electric circuit is called "Ohm's Law."

N5C02 If a current of 2 amperes flows through a 50-ohm resistor, what is the voltage across the resistor?
- A. 25 volts
- B. 52 volts
- C. 100 volts
- D. 200 volts

ANSWER C: A simple way to calculate Ohm's Law is to use the Ohm's Law magic circle:

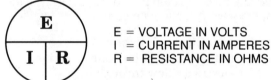

E = VOLTAGE IN VOLTS
I = CURRENT IN AMPERES
R = RESISTANCE IN OHMS

Ohm's Law Calculation

Ohm's Law ($E = I \times R$) states a relationship between voltage, current and resistance in an electrical circuit. It says that the applied electromotive force, E, in volts, is equal to the circuit current, I, in amperes, times the circuit resistance, R, in ohms. You can solve for E, I or R if the other two quantities are known. The three equations are:

$$E = I \times R \qquad I = \frac{E}{R} \qquad R = \frac{E}{I}$$

The "magic circle" helps you remember these equations. To use it, cover the unknown quantity with your finger and solve the equation for the remaining quantities. If you know the values of I and R and want to find the value of E, cover the E in the magic circle and it shows that you must multiply I times R. If you want to find I, cover the I and it shows that you must divide E by R. If you want to find R, cover the R and it shows that you must divide E by I. Since we are looking for E, the applied voltage, cover E with your finger, and you now have I (2 amps) times R (50 ohms). Multiply these two to obtain your answer of 100 volts. The calculator keystrokes are: CLEAR 2 × 50 =.

N5C03 If a 100-ohm resistor is connected to 200 volts, what is the current through the resistor?
- A. 1/2 ampere
- B. 2 amperes
- C. 300 amperes
- D. 20000 amperes

ANSWER B: In this problem, you are looking for I. Using the Ohm's Law magic circle in question N5C02, cover R with your finger. You now have E over I, or 200 over 100. Do the division, and you will end up with 2 amps. See how simple this is! Calculator keystrokes are: CLEAR 200 ÷ 100 = and your answer is 2.

N5C04 If a current of 3 amperes flows through a resistor connected to 90 volts, what is the resistance?
- A. 30 ohms
- B. 93 ohms
- C. 270 ohms
- D. 1/30 ohm

ANSWER A: Again, use the magic circle in question N5C02. In this problem, you want to find R. Covering R with your finger leaves E over I. 90 divided by 3 gives 30 ohms. See how simple it is to use Ohm's Law. Calculator keystrokes are: CLEAR 90 ÷ 3 =.

N5C05 What is the word used to describe how fast electrical energy is used?
- A. Resistance
- B. Current
- C. Power
- D. Voltage

ANSWER C: Power indicates the rate of energy being consumed.

N5C06 If you have light bulbs marked 60 watts, 75 watts and 100 watts, which one will use electrical energy the fastest?
- A. The 60 watt bulb
- B. The 75 watt bulb
- C. The 100 watt bulb
- D. They will all be the same

ANSWER C: Light bulbs that glow with different intensities consume different amounts of energy. Usually, the higher wattage bulb consumes energy faster.

N5C07 What is the basic unit of electrical power?
- A. The ohm
- B. The watt
- C. The volt
- D. The ampere

ANSWER B: Power is energy, and you all have one of those energy meters on the side of your house. You know, that's the meter that keeps turning after you've turned just about everything off! Volts times amps equals watts. There is a "magic circle" for power calculation that is similar to the one for Ohm's Law.

Here it is:

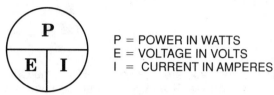

P = POWER IN WATTS
E = VOLTAGE IN VOLTS
I = CURRENT IN AMPERES

Power Calculation

As shown, P = power in watts, E = voltage in volts, and I = current in amperes. Use it in the same way as you use the Ohm's Law magic circle; that is, cover the unknown quantity with your finger and perform the mathematical operation represented by the remaining quantities. (Refer to question N5C02.)

N5C08 Which electrical circuit can have no current?
    A. A closed circuit
    B. A short circuit
    C. An open circuit
    D. A complete circuit
ANSWER C: If you turn on your equipment and nothing happens, probably something is "open." It could be that main line power switch you hid to prevent unauthorized operation.

N5C09 Which electrical circuit uses too much current?
    A. An open circuit
    B. A dead circuit
    C. A closed circuit
    D. A short circuit
ANSWER D: Anytime you have a malfunction of a piece of equipment, and you hear a pop or smell something burning, chances are a short circuit has caused the malfunction. Some electrical connection has provided a much lower resistance path for current than is normal in the circuit.

N5C10 What is the name of a current that flows only in one direction?
    A. An alternating current
    B. A direct current
    C. A normal current
    D. A smooth current
ANSWER B: Batteries generate direct current. Even though a current may vary in value, if it always flows in the same direction, it is a direct current (dc).

N5C11 What is the name of a current that flows back and forth, first in one direction, then in the opposite direction?
    A. An alternating current
    B. A direct current
    C. A rough current
    D. A reversing current
ANSWER A: If you have been shocked by house power, chances are you felt the "buzz." Be careful, it is very dangerous! Direct current (dc) flows in one direction; alternating current (ac) changes direction. Initially it flows in one direction, then it reverses and flows in the opposite direction.

N5D Concepts of frequency, including AC vs DC, frequency units, AF vs RF and wavelength

N5D01 What term means the number of times per second that an alternating current flows back and forth?
   A. Pulse rate
   B. Speed
   C. Wavelength
   D. Frequency

ANSWER D: Frequency is measured in cycles per second, called hertz. Counting the number of alternating cycles in one second gives the frequency in hertz.

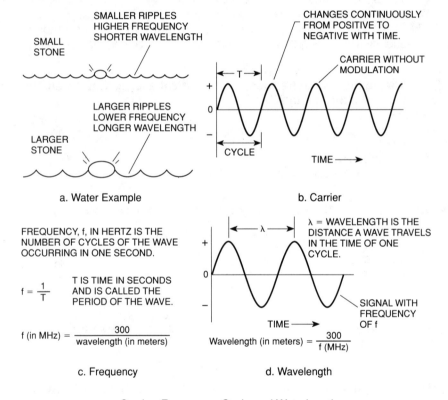

a. Water Example

b. Carrier

FREQUENCY, f, IN HERTZ IS THE NUMBER OF CYCLES OF THE WAVE OCCURRING IN ONE SECOND.

$f = \dfrac{1}{T}$ T IS TIME IN SECONDS AND IS CALLED THE PERIOD OF THE WAVE.

$f \text{ (in MHz)} = \dfrac{300}{\text{wavelength (in meters)}}$

c. Frequency

λ = WAVELENGTH IS THE DISTANCE A WAVE TRAVELS IN THE TIME OF ONE CYCLE.

SIGNAL WITH FREQUENCY OF f

$\text{Wavelength (in meters)} = \dfrac{300}{f \text{ (MHz)}}$

d. Wavelength

Carrier, Frequency, Cycle and Wavelength

N5D02 What is the basic unit of frequency?
   A. The hertz
   B. The watt
   C. The ampere
   D. The ohm

ANSWER A: The unit of frequency is named after a man named Hertz who developed horizontal antennas. When spelled out, it is written as "hertz" with a lowercase "h"; but when abbreviated, it is written as "Hz" with an uppercase "H". (Old-timers remember when the unit was cycles per second and the abbreviation was "CPS".)

N5D03 What frequency can humans hear?
   A. 0 - 20 Hz
   B. 20 - 20,000 Hz
   C. 200 - 200,000 Hz
   D. 10,000 - 30,000 Hz

ANSWER B: Your author says, "I'm not sure that I can still hear all the way up to 20,000 hertz (20 kHz). However, my dogs and cats can!"

N5D04 Why do we call signals in the range 20 Hz to 20,000 Hz audio frequencies?
   A. Because the human ear cannot sense anything in this range
   B. Because the human ear can sense sounds in this range
   C. Because this range is too low for radio energy
   D. Because the human ear can sense radio waves in this range

ANSWER B: Audio frequencies are those that you hear with your ear. Among the radio frequencies, above 20,000 hertz, are those that you can pick up with your ham receiver.

N5D05 What is the lowest frequency of electrical energy that is usually known as a radio frequency?
   A. 20 Hz
   B. 2,000 Hz
   C. 20,000 Hz
   D. 1,000,000 Hz

ANSWER C: Radio frequencies are above 20 kHz and audio frequencies are below 20 kHz.

Category	Abbrev.	Frequency	Amateur Band Wavelength
Audio	AF	20 Hz to 20 kHz	None
Very Low Frequency	VLF	3 to 30 kHz	None
Low Frequency	LF	30 to 300 kHz	None
Medium Frequency	MF	300 to 3000 kHz	160 meters
High Frequency	HF	3 to 30 MHz	80, 40, 30, 20, 17, 15, 12, 10 meters
Very High Frequency	VHF	30 to 300 MHz	6, 2, 1.25 meters
Ultrahigh Frequency	UHF	300 to 3000 MHz	70, 33, 23, 13 centimeters
Superhigh Frequency	SHF	3 to 30 GHz	9, 5, 3, 1.2 centimeters
Extremely High Frequency	EHF	Above 30 GHz	6, 4, 2.5, 2, 1 millimeter

Frequency Spectrum

N5D06 Electrical energy at a frequency of 7125 kHz is in what frequency range?
   A. Audio
   B. Radio
   C. Hyper
   D. Super-high

ANSWER B: Is 7,125,000 hertz above 20,000 hertz? You bet it is, so it must be radio frequency.

N5D07 If a radio wave makes 3,725,000 cycles in one second, what does this mean?
   A. The radio wave's voltage is 3,725 kilovolts.
   B. The radio wave's wavelength is 3,725 kilometers.

C. The radio wave's frequency is 3,725 kilohertz.

D. The radio wave's speed is 3,725 kilometers per second.

ANSWER C: 3,725 kHz is also the same thing as 3.725 MHz. We keep moving the decimal point 3 places to the left to go from Hz to kHz to MHz.

N5D08 What is the name for the distance an AC signal travels during one complete cycle?

A. Wave speed

B. Waveform

C. Wavelength

D. Wave spread

ANSWER C: A simple way to approximate a certain spot on the radio dial is to use wavelength. For instance, long-range Novice Class voice privileges are on the 10-meter band and 10 divided into 300 is 30 MHz. This approximates where you would find your Novice sub-band on 10 meters. The middle of the Novice sub-band on 10 meters is exactly 28.3 MHz, and this is frequency. Frequency is an exact spot, but the nominal wavelength for a particular band is a relatively broad area on the radio dial. See figure at question N5D01.

N5D09 What happens to a signal's wavelength as its frequency increases?

A. It gets shorter.

B. It gets longer.

C. It stays the same.

D. It disappears.

ANSWER A: The higher we go in frequency, the more cycles per second. This means wavelength gets shorter.

N5D10 What happens to a signal's frequency as its wavelength gets longer?

A. It goes down.

B. It goes up.

C. It stays the same.

D. It disappears.

ANSWER A: As wavelength gets longer, the number of cycles per second decreases. Remember, frequency and wave length are inversely proportional.

N5D11 What does 60 hertz (Hz) mean?

A. 6000 cycles per second

B. 60 cycles per second

C. 6000 meters per second

D. 60 meters per second

ANSWER B: 60 Hz means the same thing as 60 cycles per second.

Subelement N6 – Circuit Components

N6A Electrical function and/or schematic representation of resistor, switch, fuse, or battery

N6A01 What can a single-pole, double-throw switch do?
A. It can switch one input to one output.
B. It can switch one input to either of two outputs.
C. It can switch two inputs at the same time, one input to either of two outputs, and the other input to either of two outputs.
D. It can switch two inputs at the same time, one input to one output, and the other input to another output.
ANSWER B: A single-pole, double-throw switch will take a single input and switch it to either of two different outputs. The common two-position antenna switch is a good example of single-pole, double-throw.

N6A02 What can a double-pole, single-throw switch do?
A. It can switch one input to one output.
B. It can switch one input to either of two outputs.
C. It can switch two inputs at the same time, one input to either of two outputs, and the other input to either of two outputs.
D. It can switch two inputs at the same time, one input to one output, and the other input to the other output.
ANSWER D: With this switch, you may connect two contacts to two outputs.

N6A03 Which component has a positive and a negative side?
A. A battery
B. A potentiometer
C. A fuse
D. A resistor
ANSWER A: A battery has a positive and negative terminal. We usually connect a red wire to the positive terminal and a black wire to the negative terminal.

N6A04 Which component has a value that can be changed?
A. A single-cell battery
B. A potentiometer
C. A fuse
D. A resistor
ANSWER B: One type of variable resistor is called a potentiometer. We sometimes call it a "pot" for short. By turning the shaft of a "pot," we can easily change the value of resistance in a circuit.

N6A05 In Figure N6-1 which symbol represents a variable resistor or potentiometer?
A. Symbol 1
B. Symbol 2
C. Symbol 3
D. Symbol 4

ANSWER B: You can change the potential coupled to another part of a circuit with a potentiometer. The arrow in the resistor symbol indicates it represents a variable resistance.

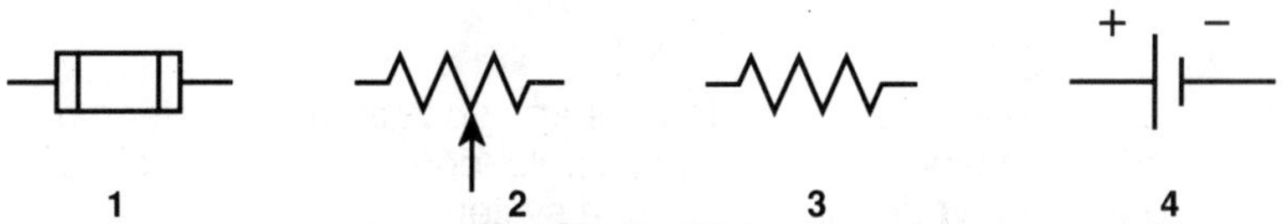

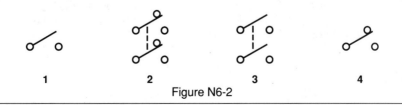

Figure N6-1

N6A06 In Figure N6-1 which symbol represents a fixed resistor?
A. Symbol 1
B. Symbol 2
C. Symbol 3
D. Symbol 4

ANSWER C: If this were a stream of water, all those bends would present resistance to the current; likewise, a resistor presents resistance to an electric current.

N6A07 In Figure N6-1 which symbol represents a fuse?
A. Symbol 1
B. Symbol 2
C. Symbol 3
D. Symbol 4

ANSWER A: This is a fuse—an intentional weak link in the circuit. You want this to melt and open the circuit in case of excessive current flow.

N6A08 In Figure N6-1 which symbol represents a single-cell battery?
A. Symbol 1
B. Symbol 2
C. Symbol 3
D. Symbol 4

ANSWER D: Battery plates never touch. The symbol looks almost like a capacitor symbol. Notice that the positive side (the long line) is indicated by a plus sign, and the negative side (the short line) is indicated by a minus sign.

N6A09 In Figure N6-2 which symbol represents a single-pole, single-throw switch?
A. Symbol 1
B. Symbol 2
C. Symbol 3
D. Symbol 4

ANSWER A: This is an on/off switch. In one position, it opens a circuit. In the other position, it closes or completes a circuit.

Figure N6-2

N6A10 In Figure N6-2 which symbol represents a single-pole, double-throw switch?
A. Symbol 1
B. Symbol 2
C. Symbol 3
D. Symbol 4

ANSWER D: If you had a double antenna system, such as a 10-meter dipole, and a 10-meter vertical, you would use a single pole, double-throw switch to allow a single radio to work off your double antenna system.

N6A11 In Figure N6-2 which symbol represents a double-pole, single-throw switch?
A. Symbol 1
B. Symbol 2
C. Symbol 3
D. Symbol 4

ANSWER C: In Figure N6-2, we see a double-pole, single-throw switch with its contacts shown in the open position.

N6A12 In Figure N6-2 which symbol represents a double-pole, double-throw switch?
A. Symbol 1
B. Symbol 2
C. Symbol 3
D. Symbol 4

ANSWER B: With this switch, you may simultaneously switch two separate signal inputs, each between two different outputs.

N6B Electrical function and/or schematic representation of a ground, antenna, transistor, or a triode vacuum tube

N6B01 Which component can amplify a small signal using low voltages?
A. A PNP transistor
B. A variable resistor
C. An electrolytic capacitor
D. A multiple-cell battery

ANSWER A: A transistor can amplify a small signal using a low-voltage power supply. (A vacuum tube amplifier requires a high-voltage power supply.)

N6B02 Which component conducts electricity from a negative emitter to a positive collector when its base voltage is made positive?
A. A variable resistor
B. An NPN transistor
C. A triode vacuum tube
D. A multiple-cell battery

ANSWER B: The transistor has an emitter, collector, and base. It will conduct electricity when operating properly. The polarities specified in the question means the transistor must be an NPN type.

N6B03 Which component is used to radiate radio energy?
A. An antenna
B. An earth ground
C. A chassis ground
D. A potentiometer
ANSWER A: It's the job of your radio station antenna to radiate your radio energy from your station's transmitter.

N6B04 In Figure N6-3 which symbol represents an earth ground?
A. Symbol 1
B. Symbol 2
C. Symbol 3
D. Symbol 4
ANSWER D: This is the classic ground symbol on worldwide radios and antenna systems. Grounding your equipment and half your antenna system is very important for safety and good range.

1   2   3   4

Figure N6-3

N6B05 In Figure N6-3 which symbol represents a chassis ground?
A. Symbol 1
B. Symbol 2
C. Symbol 3
D. Symbol 4
ANSWER A: A chassis ground is indicated by the schematic symbol illustrated in symbol 1. Chassis ground means circuit components are grounded (electrically connected) to the chassis on which they are mounted. Don't confuse it with symbol 4 which indicates earth ground.

N6B06 In Figure N6-3 which symbol represents an antenna?
A. Symbol 1
B. Symbol 2
C. Symbol 3
D. Symbol 4
ANSWER C: Looks like an antenna, doesn't it?

N6B07 In Figure N6-4 which symbol represents an NPN transistor?
A. Symbol 1
B. Symbol 2
C. Symbol 3
D. Symbol 4
ANSWER D: An easy way to identify an NPN transistor is to first identify base (B), collector (C), and emitter (E). Then look and see which way the arrow is pointing. If the arrow is NOT POINTING IN, then it's an NPN transistor.

1 2 3 4

Figure N6-4

N6B08 In Figure N6-4 which symbol represents a PNP transistor?
A. Symbol 1
B. Symbol 2
C. Symbol 3
D. Symbol 4

ANSWER A: Both symbol 1 and symbol 4 are transistors, but in symbol 1 the arrow is POINTING IN so it is a PNP transistor.

N6B09 In Figure N6-4 which symbol represents a triode vacuum tube?
A. Symbol 1
B. Symbol 2
C. Symbol 3
D. Symbol 4

ANSWER B: The triode has three elements on the inside. I know, I know, you do count four elements, but the lower element is the cathode heater, and it is usually considered as part of the cathode element. Notice that the order of the elements are in ascending alphabetical order; that is, "C" for cathode, "G" for grid, and "P" for plate. Remember that "tri" means three, like a triangle is a three-angle figure.

N6B10 What is one reason a triode vacuum tube might be used instead of a transistor in a circuit?
A. It handles higher power.
B. It uses lower voltages.
C. It uses less current.
D. It is much smaller.

ANSWER A: Transistors have replaced tubes in most radio equipment. However, the vacuum tube is still needed for high-power amplifier circuits.

N6B11 Which component can amplify a small signal but must use high voltages?
A. A transistor
B. An electrolytic capacitor
C. A vacuum tube
D. A multiple-cell battery

ANSWER C: The vacuum tube does a nice job of amplifying small signals, but tubes require relatively high voltage for their operation. (A transistor can amplify a small signal using a low-voltage power supply.)

Subelement N7 – Practical Circuits

2 exam questions
2 topic groups

N7A Functional layout of transmitter, transceiver, receiver, power supply, antenna, antenna switch, antenna feed line, impedance matching device, SWR meter

N7A01 What would you connect to your transceiver if you wanted to switch it between more than one type of antenna?
- A. A terminal-node switch
- B. An antenna switch
- C. A telegraph key switch
- D. A high-pass filter

ANSWER B: We would use an antenna switch to switch between more than one type of antenna on the roof. Stay away from inexpensive CB radio-type antenna switches—they don't work well at VHF or UHF frequencies.

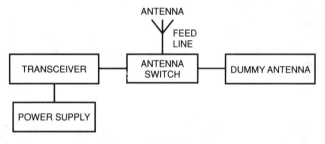

Antenna Switch Location

N7A02 What device might allow use of an antenna on a band it was not designed for?
- A. An SWR meter
- B. A low-pass filter
- C. An antenna tuner
- D. A high-pass filter

ANSWER C: We sometimes will use an antenna tuner to allow us to use an antenna on a band that it might not be specifically designed for.

N7A03 What connects your transceiver to your antenna?
- A. A dummy load
- B. A ground wire
- C. The power cord
- D. A feed line

ANSWER D: Coaxial cable is referred to as "feedline" between your transceiver and your antenna system.

N7A04 What might you connect between your transceiver and an antenna switch connected to several types of antennas?
- A. A high-pass filter
- B. An SWR meter
- C. A key click filter
- D. A mixer

ANSWER B: Use an SWR meter as a check of your antenna system and your antenna switch for a proper match. The SWR meter goes between your transceiver and the switch.

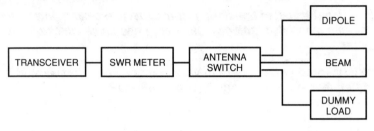

SWR Meter Location

N7A05 If your SWR meter is connected to an antenna tuner on one side, what would you connect to the other side of it?
  A. A power supply
  B. An antenna
  C. An antenna switch
  D. A transceiver

ANSWER D: The SWR meter can go between your transceiver and an antenna tuner or an antenna switch.

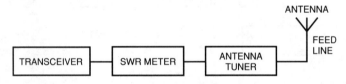

SWR Meter Between Transceiver and Antenna Tuner

N7A06 Which of these should never be connected to the output of a transceiver?
  A. An antenna switch
  B. An SWR meter
  C. An antenna
  D. A receiver

ANSWER D: Be sure to never hook up your transceiver into the antenna connection of another receiver. If you should transmit on your transceiver, you will probably damage the interconnected receiver. Never use a coax cable "T" connection to split off two sets to one antenna. Always use an antenna switch.

N7A07 If your mobile transceiver works in your car but not in your home, what should you check first?
  A. The power supply
  B. The speaker
  C. The microphone
  D. The SWR meter

ANSWER A: Most ham radio sets run off of 12 volts for mobile applications. If you plan to run your equipment in your home, you will need a power supply that converts 110 VAC house power to 12 VDC.

N7A08 What does an antenna tuner do?
A. It matches a transceiver to a mismatched antenna system.
B. It helps a receiver automatically tune in stations that are far away.
C. It switches an antenna system to a transceiver when sending, and to a receiver when listening.
D. It switches a transceiver between different kinds of antennas connected to one feed line.

ANSWER A: An antenna tuner will match your transceiver to an antenna system that might not be perfectly tuned to the frequency on which you wish to operate.

N7A09 In Figure N7-1, if block 1 is a transceiver and block 3 is a dummy antenna what is block 2?
A. A terminal-node switch
B. An antenna switch
C. A telegraph key switch
D. A high-pass filter

ANSWER B: In Figure N7-1, block 2 is an antenna switch to switch between the antenna on top of block 2 or to the dummy load which is block 3.

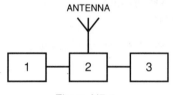

Figure N7-1

N7A10 In Figure N7-2, if block 2 is an SWR meter and block 3 is an antenna switch, what is block 1?
A. A transceiver
B. A high-pass filter
C. An antenna tuner
D. A modem

ANSWER A: Here we see a typical amateur radio installation. Block 1 is the necessary transceiver to drive block 2, the SWR meter, and block 3, the antenna switch, to switch between three antenna systems.

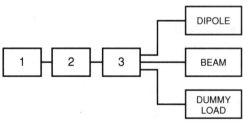

Figure N7-2

N7A11 In Figure N7-3, if block 1 is a transceiver and block 2 is an SWR meter, what is block 3?
A. An antenna switch
B. An antenna tuner

C. A key-click filter

D. A terminal-node controller

ANSWER B: There is only one antenna system in Figure N7-3, therefore, block 3 is an antenna tuner.

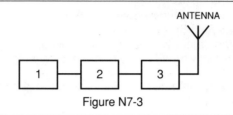

Figure N7-3

N7A12 What device converts household current to 12 VDC?

A. A catalytic converter

B. A low-pass filter

C. A power supply

D. An RS-232 interface

ANSWER C: Your hand-held transceiver may be charged at home from a power supply that converts household current to 12 VDC. Use only the charger supplied with the hand-held.

N7A13 Which of these usually needs a heavy-duty power supply?

A. An SWR meter

B. A receiver

C. A transceiver

D. An antenna switch

ANSWER C: A two-way radio transceiver requires a heavy-duty power supply. VHF and UHF equipment may require up to a 10A supply, and a big worldwide set could require up to a 20A power supply. (See figure at question N7A01.)

N7B Station layout and accessories for telegraphy, radiotelephone, radioteleprinter or packet

N7B01 What would you connect to a transceiver to send Morse code?

A. A terminal-node controller

B. A telegraph key

C. An SWR meter

D. An antenna switch

ANSWER B: Read this question carefully. Since they're asking about Morse code, you would need a telegraph key.

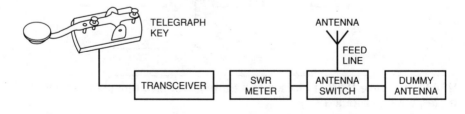

Morse Code Station

N7B02 Where would you connect a telegraph key to send Morse code?

A. To a power supply
B. To an antenna switch
C. To a transceiver
D. To an antenna

ANSWER C: The Morse code telegraph key always hooks up to a transceiver. On worldwide sets, it usually plugs in the back.

N7B03 What do many amateurs use to help form good Morse code characters?

A. A key-operated on/off switch
B. An electronic keyer
C. A key-click filter
D. A DTMF keypad

ANSWER B: If you enjoy the Morse code, you may wish to invest in an electronic keyer with side paddles. This lets you form perfect dits and dahs.

N7B04 Where would you connect a microphone for voice operation?

A. To a power supply
B. To an antenna switch
C. To a transceiver
D. To an antenna

ANSWER C: The microphone on any transceiver normally hooks up on the front of the set. Microphones for different radios usually are not interchangeable. Always stay with the mike that comes with your set. (See figure at question N7B05.)

N7B05 What would you connect to a transceiver for voice operation?

A. A splatter filter
B. A terminal-voice controller
C. A receiver audio filter
D. A microphone

ANSWER D: To transmit your voice, you need a microphone.

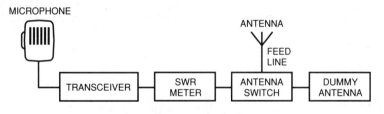

Voice or Phone Station

N7B06 What would you connect to a transceiver for RTTY operation?

A. A modem and a teleprinter or computer system
B. A computer, a printer and a RTTY refresh unit
C. A terminal voice controller
D. A modem, a monitor and a DTMF keypad

ANSWER A: If you like computers, you are going to love your new Novice Class computer privileges! Chances are you already own most of the components for going on radioteleprinter, or packet operation. Your present personal computer and printer may need only to be connected to a modem or terminal-node controller (TNC). These easily plug into your ham radio set, and you are on the air! (See figure at question N7B07.)

N7B07 What would you connect between a transceiver and a computer system or teleprinter for RTTY operation?

A. An RS-232 interface
B. A DTMF keypad
C. A modem
D. A terminal-network controller

ANSWER C: As you can see, it's not complicated to tie in your ham set for digital communications. Many manufacturers offer transceiver tie-in kits that allow you to interface your set and your computer without ever having to pick up a screwdriver! Be careful here because Answer D says "terminal-network controller" not "terminal-node controller."

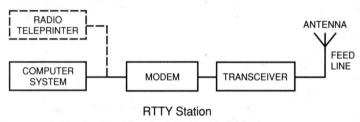

RTTY Station

N7B08 What would you connect between a computer system and a transceiver for packet-radio operation?

A. A terminal-node controller
B. A DTMF keypad
C. An SWR bridge
D. An antenna tuner

ANSWER A: The heart of your packet station will be a microprocessor-based device called a terminal-node (not network) controller (TNC). It allows you to use your computer on the ham airwaves. Its function is to assemble and disassemble the packets of information. There may be several models available for any type of worldwide or VHF/UHF ham set. If you live on a mountain top, your TNC can also turn your station into a digipeater for the automatic relay of packet transmissions.

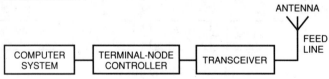

Packet-Radio Station

N7B09 Where would you connect a terminal-node controller for packet-radio operation?

A. Between your antenna and transceiver
B. Between your computer and monitor
C. Between your computer and transceiver
D. Between your keyboard and computer

ANSWER C: The terminal-node controller interfaces your computer to your ham set so that you can automatically receive computer messages via Amateur Radio. Most controllers just plug into your computer without any soldering or tools. (See figure at question N7B08.)

N7B10 In RTTY operation, what equipment connects to a modem?
   A. A DTMF keypad, a monitor and a transceiver
   B. A DTMF microphone, a monitor and a transceiver
   C. A transceiver and a terminal-network controller
   D. A transceiver and a teleprinter or computer system
ANSWER D: If you plan to operate RTTY, you will need a transceiver and a radioteleprinter or computer system. (See figure at question N7B07.)

N7B11 In packet-radio operation, what equipment connects to a terminal-node controller?
   A. A transceiver and a modem
   B. A transceiver and a terminal or computer system
   C. A DTMF keypad, a monitor and a transceiver
   D. A DTMF microphone, a monitor and a transceiver
ANSWER B: It's important to tell the salesperson the type of ham set and type of computer you have in order to get the right TNC and interconnecting cables. (See figure at question N7B08.)

Subelement N8 – Signals and Emissions	2 exam questions 2 topic groups

N8A Emission types, key clicks, chirps or superimposed hum

N8A01 How is CW usually transmitted?
   A. By frequency-shift keying an RF signal
   B. By on/off keying an RF signal
   C. By audio-frequency-shift keying an oscillator tone
   D. By on/off keying an audio-frequency signal
ANSWER B: The telegraph key turns on and off the radio frequency signal. CW is Morse code (earlier hams know it as A1A). This signal has no modulation—it's just interrupted carrier CW (continuous wave). The carrier is on for the duration of a "dit" or "dah" and is off the rest of the time. See figure at question N8A03.

N8A02 How is RTTY usually transmitted?
   A. By frequency-shift keying an RF signal
   B. By on/off keying an RF signal
   C. By digital pulse-code keying of an unmodulated carrier
   D. By on/off keying an audio-frequency signal
ANSWER A: This is many times abbreviated FSK, and it's radioteletype. Here the frequency of the carrier is shifted to identify the information being transmitted.

N8A03 What is the name for international Morse code emissions?
   A. RTTY
   B. Data
   C. CW
   D. Phone

ANSWER C: CW stands for continuous wave, another name for Morse code. See the emission definitions at question N1E01. As shown in the figure, the continuous wave is interrupted for sending code. As stated previously, earlier hams know CW as A1A.

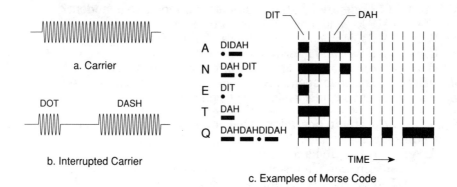

a. Carrier

b. Interrupted Carrier

c. Examples of Morse Code

Transmitting CW

N8A04 What is the name for narrow-band direct-printing telegraphy emissions?
  A. RTTY
  B. Data
  C. CW
  D. Phone
ANSWER A: Radioteleprinting is a popular Amateur Radio pastime. RTTY stands for radioteletype. See the emission definitions at question N1E01. Earlier hams know this emission as F1B.

N8A05 What is the name for packet-radio emissions?
  A. RTTY
  B. Data
  C. CW
  D. Phone
ANSWER B: The digital data stream of computer communications is called packet.

N8A06 What is the name for voice emissions?
  A. RTTY
  B. Data
  C. CW
  D. Phone
ANSWER D: Phone (or radiotelephone) is another name for voice communications. See the emission definitions at question N1E01.

N8A07 How can you prevent key clicks?
  A. By sending CW more slowly
  B. By increasing power
  C. By using a better power supply
  D. By using a key-click filter

ANSWER D: Unless you build your own gear, you shouldn't have a problem with key clicks. Modern transceivers for worldwide transmission use have a built-in key-click filter. All you need to do is to buy a key, plug it in, and have fun with CW.

N8A08 What does chirp mean?

A. An overload in a receiver's audio circuit whenever CW is received

B. A high-pitched tone which is received along with a CW signal

C. A small change in a transmitter's frequency each time it is keyed

D. A slow change in transmitter frequency as the circuit warms up

ANSWER C: Chirps are changes in frequency at the beginning of a keying pulse and are usually caused by poor power supply voltage regulation. Many worldwide ham sets operate only from 12 volts. This keeps their size down. For home use, you will need at least a 20-amp power supply to run your new worldwide 10-meter setup. If you try to use an old CB-type power supply, chances are it will pull the voltage down so low that your transmitter's keyed tone will wobble, or chirp, in frequency.

N8A09 What can be done to keep a CW transmitter from chirping?

A. Add a low-pass filter.

B. Use an RF amplifier.

C. Keep the power supply current very steady.

D. Keep the power supply voltages very steady.

ANSWER D: A signal with chirp is an unstable signal. Unstable low voltage is many times the culprit. A better regulation of the power supply output voltage will cure the problem. If you are running on batteries, charge the batteries!

N8A10 What may cause a buzzing or hum in the signal of an HF transmitter?

A. Using an antenna which is the wrong length

B. Energy from another transmitter

C. Bad design of the transmitter's RF power output circuit

D. A bad filter capacitor in the transmitter's power supply

ANSWER D: If you find a 20-year-old power supply your dad used back in the good old days for his ham radio setup, better check its dc output power. Very old power supplies feature big electrolytic filter capacitors that dry out and lose their capacity with age. The result will be severe hum on transmit and receive. Power supply technology has changed so rapidly that switching power supplies can easily deliver 12 volts at 20 amps in small and relatively lightweight packages. They have done away with the very heavy transformers needed in the older conventional power supplies. The switching power supply also doesn't have those big filter capacitors that dry out.

N8A11 Which sideband is commonly used for 10-meter phone operation?

A. Upper-sideband

B. Lower-sideband

C. Amplitude-compandored sideband

D. Double-sideband

ANSWER A: We use upper sideband (USB) on 10 meters, 12 meters, 15 meters, and 20 meters. We operate lower sideband (LSB) on 40 meters, 80 meters, and 160 meters.

N8B Harmonics and unwanted signals, equipment and adjustments to help reduce interference to others

N8B01 How does the frequency of a harmonic compare to the desired transmitting frequency?
A. It is slightly more than the desired frequency.
B. It is slightly less than the desired frequency.
C. It is exactly two, or three, or more times the desired frequency.
D. It is much less than the desired frequency.

ANSWER C: Harmonics are multiples of your fundamental frequency, and are not desirable.

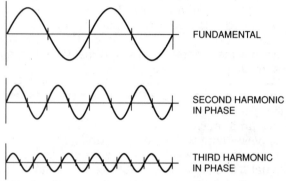

Fundamental Radio Wave and Harmonics

N8B02 What is the fourth harmonic of a 7160-kHz signal?
A. 28,640 kHz
B. 35,800 kHz
C. 28,160 kHz
D. 1790 kHz

ANSWER A: Harmonics are multiples of your transmitted signal. Simply multiply the harmonic number times the fundamental frequency. In this case, multiply 4 times 7160 kHz and the result is Answer A. Calculator keystrokes are: CLEAR 4 × 7160 = and the answer is 28640.

N8B03 If you are told your station was heard on 21,375 kHz, but at the time you were operating on 7125 kHz, what is one reason this could happen?
A. Your transmitter's power-supply filter capacitor was bad
B. You were sending CW too fast
C. Your transmitter was radiating harmonic signals
D. Your transmitter's power-supply filter choke was bad

ANSWER C: Because these two frequencies are harmonically related (3 × 7125 kHz = 21,375 kHz), chances are your ham set is an older model and is generating harmonics. You need to work on the radio, or get it fixed professionally. Newer ham sets rarely have this problem.

N8B04 If someone tells you that signals from your hand-held transceiver are interfering with other signals on a frequency near yours, what may be the cause?

A. You may need a power amplifier for your hand-held.

B. Your hand-held may have chirp from weak batteries.

C. You may need to turn the volume up on your hand-held.

D. Your hand-held may be transmitting spurious emissions.

ANSWER D: If you receive a report from an amateur operator at least 5 miles away that your signals are interfering with other signals on other frequencies, your hand-held may be transmitting spurious emissions. You will need to take your set in to an authorized equipment dealer for a quick checkout.

N8B05 If your transmitter sends signals outside the band where it is transmitting, what is this called?

A. Off-frequency emissions

B. Transmitter chirping

C. Side tones

D. Spurious emissions

ANSWER D: Here again the question says "signals," so they are surely spurious emissions.

N8B06 What problem may occur if your transmitter is operated without the cover and other shielding in place?

A. It may transmit spurious emissions.

B. It may transmit a chirpy signal.

C. It may transmit a weak signal.

D. It may interfere with other stations operating near its frequency.

ANSWER A: If you take your transmitter apart, make sure you completely reassemble it. If you leave out any metal covers, chances are it could radiate spurious emissions.

N8B07 What may happen if an SSB transmitter is operated with the microphone gain set too high?

A. It may cause digital interference to computer equipment.

B. It may cause splatter interference to other stations operating near its frequency.

C. It may cause atmospheric interference in the air around the antenna.

D. It may cause interference to other stations operating on a higher frequency band.

ANSWER B: All worldwide transceivers have a microphone gain control. Many base station microphones also offer audio compression and additional "mike" gain. Turning up the mike gain too high will cause distortion, splatter, and interference to stations on nearby frequencies. It's also bad Amateur Radio practice to operate with a "hot" mike setting. Read your instruction manual for proper microphone gain setting procedures while watching the ALC level on your transceiver's multimeter.

N8B08 What may happen if an SSB transmitter is operated with too much speech processing?

A. It may cause digital interference to computer equipment.

B. It may cause splatter interference to other stations operating near its frequency.

C. It may cause atmospheric interference in the air around the antenna.

D. It may cause interference to other stations operating on a higher frequency band.

ANSWER B: Modern worldwide transceivers, such as the set you will use on 10 meters, contain powerful speech processing circuits that intensify your transmitted voice signal. Too high a setting of the speech processor will lead to splatter and terrible sounding transmissions. In fact, avoid using the speech processor unless absolutely necessary. Using the speech processor widens your transmitted signal, possibly causing interference to stations operating on frequencies close to yours.

N8B09 What may happen if an FM transmitter is operated with the microphone gain or deviation control set too high?

A. It may cause digital interference to computer equipment.

B. It may cause interference to other stations operating near its frequency.

C. It may cause atmospheric interference in the air around the antenna.

D. It may cause interference to other stations operating on a higher frequency band.

ANSWER B: On your FM equipment, there is no microphone gain control. It has been preset at the factory, so you probably won't ever need to worry about microphone gain when operating with a stock microphone. However, if you decide to change to a different type of microphone, you may wish to let a ham radio expert check the audio level.

N8B10 What may your FM hand-held or mobile transceiver do if you shout into its microphone?

A. It may cause digital interference to computer equipment.

B. It may cause interference to other stations operating near its frequency.

C. It may cause atmospheric interference in the air around the antenna.

D. It may cause interference to other stations operating on a higher frequency band.

ANSWER B: Never shout into the microphone—it may cause interference to other stations on adjacent frequencies. Speak in a normal tone of voice with the microphone held approximately one inch away from your mouth.

N8B11 What can you do if you are told your FM hand-held or mobile transceiver is over deviating?

A. Talk louder into the microphone.

B. Let the transceiver cool off.

C. Change to a higher power level.

D. Talk farther away from the microphone.

ANSWER D: If your set is over-deviating, it means that too much modulation is driving your signal beyond its normal bandwidth. If you talk farther away from the microphone, you will temporarily minimize over-deviation.

3 exam questions
3 topic groups

N9A Wavelength vs antenna length

N9A01 How do you calculate the length (in feet) of a half-wavelength dipole antenna?
    A. Divide 150 by the antenna's operating frequency (in MHz) [150/f(in MHz)]
    B. Divide 234 by the antenna's operating frequency (in MHz) [234/f (in MHz)]
    C. Divide 300 by the antenna's operating frequency (in MHz) [300/f (in MHz)]
    D. Divide 468 by the antenna's operating frequency (in MHz) [468/f (in MHz)]
ANSWER D: Plug this into your memory—you will use it often in your ham radio career. (See question N9A03.)

N9A02 How do you calculate the length (in feet) of a quarter-wavelength vertical antenna?
    A. Divide 150 by the antenna's operating frequency (in MHz) [150/f (in MHz)]
    B. Divide 234 by the antenna's operating frequency (in MHz) [234/f (in MHz)]
    C. Divide 300 by the antenna's operating frequency (in MHz) [300/f (in MHz)]
    D. Divide 468 by the antenna's operating frequency (in MHz) [468/f (in MHz)]
ANSWER B: A quarterwave vertical antenna will be half as long as a half-wave dipole. For a half-wave dipole, as shown in question N9A03, divide 468 by the antenna's operating frequency in MHz. Therefore, for a quarter-wave antenna, divide one-half of 468, or 234, by the antenna's operating frequency in MHz.

N9A03 If you made a half-wavelength dipole antenna for 3725 kHz, how long would it be (to the nearest foot)?
    A. 126 ft
    B. 81 ft
    C. 63 ft
    D. 40 ft
ANSWER A: A simple antenna to build for worldwide operation is called the half-wave dipole. This simple antenna contains everyday wire whose length is one-half of the wavelength of the frequency being transmitted. The signal from your transmitter is applied in the exact middle of the dipole. It is best to use coaxial cable for this feed-in connection. The length of a half-wave dipole is easily calculated using the equation:

$$l = \frac{468}{f}$$

where l is the wavelength in feet and f is the frequency in MHz. To use the equation, you must convert kilohertz to megahertz. Move the decimal point three places to the left to convert 3725 kHz to 3.725 MHz. Now divide 3.725

into 468. The calculator keystrokes are: CLEAR 468 ÷ 3.725 = and the answer is 125.6 feet. Round it to 126 feet to match the answer choice. Try it, it's easy! Remember 468 to calculate the length in feet of a half-wave dipole.

N9A04 If you made a half-wavelength dipole antenna for 28.150 MHz, how long would it be (to the nearest foot)?
  A. 22 ft
  B. 11 ft
  C. 17 ft
  D. 34 ft

ANSWER C: Here we are at the 10-meter band, 28 MHz. This is where your new Novice Class privileges allow voice, so this dipole will be quite popular. There is good news; after you calculate its length (see question N9A03), you will see that it's only 17 feet long. With this antenna you can go almost anywhere! Calculator keystrokes are: CLEAR 468 ÷ 28.15 = and the answer is 16.6. Round it to 17 to match the answer choice.

N9A05 If you made a quarter-wavelength vertical antenna for 7125 kHz, how long would it be (to the nearest foot)?
  A. 11 ft
  B. 16 ft
  C. 21 ft
  D. 33 ft

ANSWER D: This question is for a quarter-wave antenna as in question N9A02. The higher you go in frequency, the shorter the dipole. Remember to change kilohertz to megahertz (7.125 MHz) and divide the result into 234, not 468. Calculator keystrokes are: CLEAR 234 ÷ 7.125 = and the answer is 32.8. Round it to 33 to match the answer choice.

N9A06 If you made a quarter-wavelength vertical antenna for 21.125 MHz, how long would it be (to the nearest foot)?
  A. 7 ft
  B. 11 ft
  C. 14 ft
  D. 22 ft

ANSWER B: Here, as in question N9A02, you are being asked for the length of a quarter-wavelength antenna, not a half-wavelength. Again, the higher you go in frequency, the shorter the dipole length. Convert kilohertz to megahertz by moving the decimal point three places to the left (21.125 MHz), and divide into 234. Calculator keystrokes are: CLEAR 234 ÷ 21.125 = and the answer is 11.1. Round it to 11 to match the answer choice.

N9A07 If you made a half-wavelength vertical antenna for 223 MHz, how long would it be (to the nearest inch)?
  A. 112 inches
  B. 50 inches
  C. 25 inches
  D. 12 inches

ANSWER C: Use the formula 468 divided by the frequency in MHz for a half-wavelength antenna in feet. You must multiply your answer by 12 to arrive at 25 inches.

N9A08 If an antenna is made longer, what happens to its resonant frequency?
A. It decreases.
B. It increases.
C. It stays the same.
D. It disappears.
ANSWER A: Remember the rule: "lower-longer." If an antenna is cut a little bit too long, it works better at a lower frequency.

N9A09 If an antenna is made shorter, what happens to its resonant frequency?
A. It decreases.
B. It increases.
C. It stays the same.
D. It disappears.
ANSWER B: If you trim the tip of a vertical antenna to make it shorter, its resonant frequency will increase.

N9A10 How could you lower the resonant frequency of a dipole antenna?
A. Lengthen the antenna
B. Shorten the antenna
C. Use less feed line
D. Use a smaller size feed line
ANSWER A: To lower the resonant frequency of an antenna, you will need to add wire to the antenna to make it longer.

N9A11 How could you raise the resonant frequency of a dipole antenna?
A. Lengthen the antenna.
B. Shorten the antenna.
C. Use more feed line.
D. Use a larger size feed line.
ANSWER B: To raise the resonant frequency of a dipole antenna, slightly shorten each end.

N9B Yagi parts, concept of directional antennas, and safety near antennas

N9B01 In what direction does a Yagi antenna send out radio energy?
A. It goes out equally in all directions.
B. Most of it goes in one direction.
C. Most of it goes equally in two opposite directions.
D. Most of it is aimed high into the air.
ANSWER B: Dr. H. Yagi, a Japanese physicist, translated into English the directional beam antenna design named after him.

Actually, a professor named Uda invented the antenna with the placement of a slightly shorter dipole in front of, and slightly longer dipole in back of, a regular dipole to cause the signal to concentrate in one direction. The longer element reflected some of the energy and the shorter element directed the energy in one specific direction. So the Yagi antenna is really a series of slightly longer and shorter dipoles that influence a directly fed dipole to transmit better in a forward direction. (See Figure N9-1.)

N9B02 About how long is the driven element of a Yagi antenna?
A. 1/4 wavelength
B. 1/3 wavelength
C. 1/2 wavelength
D. 1 wavelength

ANSWER C: Since the Yagi antenna is a series of dipoles affixed to a boom in the same plane, the individual elements are about one-half wavelength long, just like the driven element. The reflector is a little longer, the director a little shorter. (See Figure N9-1.)

N9B03 In Figure N9-1, what is the name of element 2 of the Yagi antenna?
A. Director
B. Reflector
C. Boom
D. Driven element

ANSWER D: The coaxial cable always connects to the driven element. It is driven by the direct energy from the transmitter.

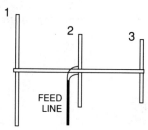

Figure N9-1

N9B04 In Figure N9-1, what is the name of element 3 of the Yagi antenna?
A. Director
B. Reflector
C. Boom
D. Driven element

ANSWER A: The director is usually the shortest element and is placed in front of the driven element. Directors help concentrate the signal into a tight radiation pattern.

N9B05 In Figure N9-1, what is the name of element 1 of the Yagi antenna?
A. Director
B. Reflector
C. Boom
D. Driven element

ANSWER B: The reflector is on the back of the beam. A reflector is always longer than the driven element. Sometimes there are two or more reflectors.

N9B06 Looking at the Yagi antenna in Figure N9-1, in which direction on the page would it send most of its radio energy?
A. Left
B. Right
C. Top
D. Bottom

ANSWER B: The Yagi in Figure N9-1 has the shortest end—the director—facing to the right. This means it will radiate most of its energy to the right as you look at it on the paper.

N9B07 Why is a 5/8-wavelength vertical antenna better than a 1/4-wavelength vertical antenna for VHF or UHF mobile operations?
    A. A 5/8-wavelength antenna can handle more power.
    B. A 5/8-wavelength antenna has more gain.
    C. A 5/8-wavelength antenna has less corona loss.
    D. A 5/8-wavelength antenna is easier to install on a car.
ANSWER B: There are about ten manufacturers of mobile VHF and UHF ham antennas. Some are trunk-lip mounted, some go through the car chassis (ugh!), and some will even hang onto the window glass. A 5/8-wave antenna is much taller than the quarterwave antenna, and it has more gain.

N9B08 In what direction does a vertical antenna send out radio energy?
    A. Most of it goes in two opposite directions.
    B. Most of it goes high into the air.
    C. Most of it goes equally in all horizontal directions.
    D. Most of it goes in one direction.
ANSWER C: A dipole antenna radiates the signal bidirectionally—broadside to the wire. A vertical antenna radiates the signal vertically equally in all horizontal directions.

N9B09 If the ends of a half-wave dipole antenna point east and west, which way would the antenna send out radio energy?
    A. Equally in all directions
    B. Mostly up and down
    C. Mostly north and south
    D. Mostly east and west
ANSWER C: If the dipole is erected east and west, the energy would go out mostly north and south. Slightly droop the dipole ends if you wish more energy in other directions.

N9B10 How should you hold the antenna of a hand-held transceiver while you are transmitting?
    A. Away from your head and away from others
    B. Pointed towards the station you are contacting
    C. Pointed away from the station you are contacting
    D. Pointed down to bounce the signal off the ground
ANSWER A: Even little hand-held radios could pose a radiation hazard if the antenna is too close to your eyes. The key word here is "hand-held."

N9B11 Why should your outside antennas be high enough so that no one can touch them while you are transmitting?
    A. Touching the antenna might cause television interference
    B. Touching the antenna might cause RF burns
    C. Touching the antenna might radiate harmonics
    D. Touching the antenna might reflect the signal back to the transmitter and cause damage
ANSWER B: It makes good sense to always keep antennas away from someone that might touch them.

N9C *Feed lines, baluns and polarization via element orientation*

N9C01 What is a coaxial cable?
A. Two wires side-by-side in a plastic ribbon
B. Two wires side-by-side held apart by insulating rods
C. Two wires twisted around each other in a spiral
D. A center wire inside an insulating material covered by a metal sleeve or shield

ANSWER D: Coaxial cable is similar to water pipes in that it doesn't leak and carries the good stuff inside while keeping out everything else. To maintain high water pressure at the output end requires large diameter pipes. Same thing with coaxial cable—if you plan to run a lot of power, or if you have an extremely long cable run, use a large diameter coaxial cable.

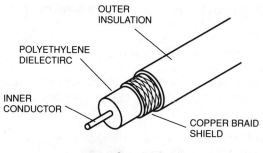

Coaxial Cable

N9C02 Why does coaxial cable make a good antenna feed line?
A. You can make it at home, and its impedance matches most amateur antennas.
B. It is weatherproof, and its impedance matches most amateur antennas.
C. It is weatherproof, and its impedance is higher than that of most amateur antennas.
D. It can be used near metal objects, and its impedance is higher than that of most amateur antennas.

ANSWER B: You can even bury non-contaminating quality coaxial cable. Your author's station uses coaxial cable exclusively, and external runs are underground in PVC tubes so the gophers can't get to the cable.

N9C03 Which kind of antenna feed line can carry radio energy very well even if it is buried in the ground?
A. Twin lead
B. Coaxial cable
C. Parallel conductor
D. Twisted pair

ANSWER B: It's very important to completely weatherproof the antenna connection that is out in the open. Coaxial cable connectors may leak moisture, so seal them well.

N9C04 What is the best antenna feed line to use if it must be put near grounded metal objects?

A. Coaxial cable
B. Twin lead
C. Twisted pair
D. Ladder-line

ANSWER A: Since the shield is at ground potential, running coaxial cable close to grounded objects will not affect the signal. If you need to hide your coaxial cable on a run from the ground floor to the roof, consider running it inside a rain gutter's down-spout.

N9C05 What is parallel-conductor feed line?
A. Two wires twisted around each other in a spiral
B. Two wires side-by-side held apart by insulating rods
C. A center wire inside an insulating material which is covered by a metal sleeve or shield
D. A metal pipe which is as wide or slightly wider than a wavelength of the signal it carries

ANSWER B: A less-used feedline is "ladder line." It's similar to TV twin-lead. Ladder line can be used with certain types of wire antenna systems. Ladder line cannot be connected directly to any ham set; it must have an additional antenna tuning device.

INSULATOR   PARALLEL-CONDUCTOR "LADDER LINE"   TWIN LEAD

Parallel Conductor and Twin Lead

N9C06 What are some reasons to use parallel-conductor feed line?
A. It has low impedance, and will operate with a high SWR.
B. It will operate with a high SWR, and it works well when tied down to metal objects.
C. It has a low impedance, and has less loss than coaxial cable.
D. It will operate with a high SWR, and has less loss than coaxial cable.

ANSWER D: There was a ham that lived in a valley who put a wire antenna several thousand feet away on the top of a hill. He used parallel conductor feedline to connect his set with the hilltop antenna because there is less loss in this type of open-wire feed system than in coaxial cable.

N9C07 What are some reasons not to use parallel-conductor feed line?
A. It does not work well when tied down to metal objects, and you must use an impedance matching device with your transceiver.
B. It is difficult to make at home, and it does not work very well with a high SWR.
C. It does not work well when tied down to metal objects, and it cannot operate under high power.
D. You must use an impedance matching device with your transceiver, and it does not work very well with a high SWR.

ANSWER A: Parallel-conductor feed line cannot be run beside any metal object. Also, you'll need an antenna tuner to transform its 300-ohm to 600-ohm impedance to your transceiver's 52-ohm impedance.

N9C08 What kind of antenna feed line is made of two conductors held apart by insulated rods?
A. Coaxial cable
B. Open-conductor ladder line
C. Twin lead in a plastic ribbon
D. Twisted pair
ANSWER B: It is sometimes called a parallel open-wire feedline. While not very popular with today's amateurs, it can be built by using solid copper wire and plastic spacers.

N9C09 What would you use to connect a coaxial cable of 50-ohm impedance to an antenna of 35-ohm impedance?
A. A terminating resistor
B. An SWR meter
C. An impedance matching device
D. A low-pass filter
ANSWER C: Many homemade vertical antennas will have an impedance mismatch. We use a small impedance matching device to match the antenna precisely to the transceiver.

N9C10 What does balun mean?
A. Balanced antenna network
B. Balanced unloader
C. Balanced unmodulator
D. Balanced to unbalanced
ANSWER D: We use a balun to match balanced twin-lead to unbalanced 50-ohm coaxial cable. Balun receives its name from balanced to unbalanced. It is an impedance-matching transformer.

N9C11 Where would you install a balun to feed a dipole antenna with 50-ohm coaxial cable?
A. Between the coaxial cable and the antenna
B. Between the transmitter and the coaxial cable
C. Between the antenna and the ground
D. Between the coaxial cable and the ground
ANSWER A: This is a much more common scenario—coaxial cable feeding a balanced dipole. In this case, the balun is installed on the antenna at the antenna feedpoint.

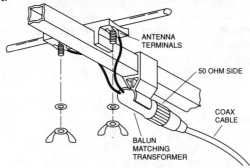

Using Balun to Feed Dipole

Source: *Using Video In Your Home*, G. McComb, © 1989 Master Publishing, Inc.

CHECK YOURSELF

At this point, go back and check your work. Continue to review the questions and answers until you have mastered them so you don't miss any more than 1 in 10. Remember, you will have a 30-question written examination on Element 2, and you must get 22 correct answers. Have someone give you a trial examination to see how you do.

Element 3A Question Pool

When you are confident that you have mastered Element 2, continue the same study pattern with Element 3A. Here are the 295 Element 3A questions, the correct answers, and an explanation for each correct answer. In addition, the syllabus used to develop the question pool is included for your guidance. Continue to study hard, and good luck!

Element 3A (Technician Class) Syllabus

Here is the syllabus used to develop the question pool:

T1 – Commission's Rules
(5 exam questions – 5 groups)
T1A Station control, frequency privileges authorized to the Technician class control operator, term of licenses, grace periods and modifications of licenses
T1B Emission privileges for Technician class control operator, frequency selection and sharing, transmitter power
T1C Digital communications, station identification, ID with CSCE
T1D Correct language, Phonetics, Beacons and Radio Control of model craft and vehicles
T1E Emergency communications; braodcasting; permissible one-way, satellite and third party communication; indecent and profane language

T2 – Operating Procedures
(3 exam questions – 3 groups)
T2A Repeater operation, courteous operation
T2B Simplex operations, Q signals, RST signal reporting, repeater frequency coordination
T2C Distress calling and emergency drills and communications - operations and equipment, Radio Amateur Civil Emergency Service (RACES)

T3 – Radio-Wave Propagation
(3 exam questions – 3 groups)
T3A Ionosphere, Ionospheric regions, solar radiation
T3B Ionospheric absorption, causes and variation, maximum usable frequency
T3C Propagation, including ionospheric, tropospheric, line-of-sight scatter propagation, and Maximum Usable Frequency

T4 – Amateur Radio Practices
(4 exam questions – 4 groups)
T4A Electrical wiring, including switch location, dangerous voltages and currents
T4B Meters, including volt, amp, multi, peak-

reading, RF watt and placement, and ratings of fuses and switches
T4C Marker generator, crystal calibrator, signal generators and impedance-match indicator
T4D Dummy antennas, S-meter, exposure of the human body to RF

T5 – Electrical Principles
(2 exam questions – 2 groups)
T5A Definition of resistance, inductance, and capacitance and unit of measurement, calculation of values in series and parallel
T5B Ohm's Law

T6 – Circuit Components
(2 exam questions – 2 groups)
T6A Resistors, construction types, variable and fixed, color code, power ratings, schematic symbols
T6B Schematic symbols - inductors and capacitors, construction of variable and fixed, factors affecting inductance and capacitance, capacitor consturction

T7 – Practical Circuits
(1 exam question – 1 group)
T7A Practical circuits

T8 – Signals and Emission
(2 exam questions – 2 groups)
T8A Definition of modulation and emission types
T8B RF carrier, modulation, bandwidth and deviation

T9 – Antennas and Feed Lines
(3 exam questions – 3 groups)
T9A Parasitic beam and non-directional antennas
T9B Polarization, impedance matching and SWR, feed lines, balanced vs unbalanced (including baluns)
T9C Line losses by line type, length and frequency, RF safety

Subelement T1 – Commission's Rules

Note: A §Part 97 reference is enclosed in brackets, e.g., [97], after each correct answer explanation in this subelement.

T1A Station control, frequency privileges authorized to the Technician Class control operator, term of licenses, grace periods and modifications of licenses

T1A01 What is the control point of an amateur station?
A. The on/off switch of the transmitter
B. The input/output port of a packet controller
C. The variable frequency oscillator of a transmitter
D. The location at which the control operator function is performed
ANSWER D: This is where you have complete capabilities to turn the equipment on, or shut it off, in case of a malfunction. Every ham radio station is required to have a control point. [97.3a12]

T1A02 What is the term for the location at which the control operator function is performed?
A. The operating desk
B. The control point
C. The station location
D. The manual control location
ANSWER B: Some repeaters that may be used for phone patches are controlled by radio links. This means that the actual control point is with the control operator and his tiny hand-held transceiver. When you upgrade to Technician, you too could be a control operator with the control point worn on your belt! [97.3a12]

T1A03 What must you do to renew or change your operator/primary station license?
A. Properly fill out FCC Form 610 and send it to the FCC in Gettysburg, PA.
B. Properly fill out FCC Form 610 and send it to the nearest FCC field office.
C. Properly fill out FCC form 610 and send it to the FCC in Washington, DC.
D. An amateur license never needs changing or renewing.
ANSWER A: Don't write a letter! All modifications and renewals are done on FCC Form 610. There is a Form 610 bound into the back of this book. [97.19a/b]

T1A04 What is the "grace period" during which the FCC will renew an expired 10-year license?
A. 2 years
B. 5 years
C. 10 years
D. There is no grace period

ANSWER A: You are not allowed to operate during a grace period; however, you can keep your privileges for 2 years. After that, they are lost for good. So is your call sign. Don't forget to renew! [97.19c]

T1A05 Which of the following frequencies may a Technician operator who has passed a Morse code test use?
A. 7.1 - 7.2 MHz
B. 14.1 - 14.2 MHz
C. 21.1 - 21.2 MHz
D. 28.1 - 29.2 MHz

ANSWER C: 21.1-21.2 MHz is the upper and lower limits to the 15- meter Novice band. Watch out for Answers A and D—they start out with the correct frequency, but they end up wrong. Answer C is correct. [97.301/305e]

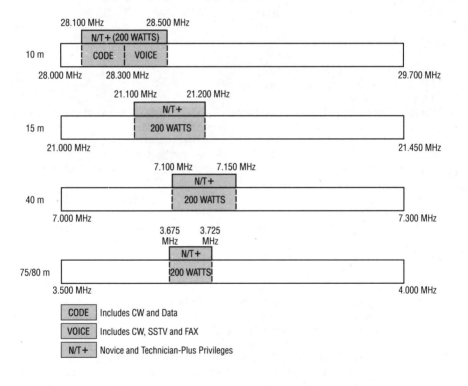

Technician-Plus Class Privileges on HF Bands

T1A06 Which operator licenses authorize privileges on 52.525 MHz?
A. Extra, Advanced only
B. Extra, Advanced, General only
C. Extra, Advanced, General, Technician only
D. Extra, Advanced, General, Technician, Novice

ANSWER C: 52.525 MHz is the middle of the 6-meter wavelength band. Every-one but the Novice has privileges here. [97.301a]

T1A07 Which operator licenses authorize privileges on 146.52 MHz?
  A. Extra, Advanced, General, Technician, Novice
  B. Extra, Advanced, General, Technician only
  C. Extra, Advanced, General only
  D. Extra, Advanced only

ANSWER B: 146.52 MHz is in the middle of the 2-meter wavelength band. Again, everybody but the Novice has privileges here. [97.301a]

T1A08 Which operator licenses authorize privileges on 223.50 MHz?
  A. Extra, Advanced, General, Technician, Novice
  B. Extra, Advanced, General, Technician only
  C. Extra, Advanced, General only
  D. Extra, Advanced only

ANSWER A: This is in the 1.25 m Novice band, and since Novices have voice privileges up here, everyone is allowed on this frequency. [97.301a]

T1A09 Which operator licenses authorize privileges on 446.0 MHz?
  A. Extra, Advanced, General, Technician, Novice
  B. Extra, Advanced, General, Technician only
  C. Extra, Advanced, General only
  D. Extra, Advanced only

ANSWER B: The 440-MHz band is off limits to Novices—but open to everyone else. 446.0 MHz is a fun spot for UHF operation. [97.301a]

T1A10 In addition to passing the Technician written examination (Elements 2 and 3A), what must you do before you are allowed to use amateur frequencies below 30 MHz?
  A. Nothing special is needed; all Technicians may use the HF bands at any time.
  B. You must notify the FCC that you intend to operate on the HF bands.
  C. You must attend a class to learn about HF communications.
  D. You must pass a Morse code test (either Element 1A, 1B or 1C).

ANSWER D: As a no-code Technician operator, you do not have Novice code privileges until you pass the 5-wpm code test in front of a team of three accredited volunteer examiners. You could also satisfy the code requirement by passing the 13-wpm or the 20-wpm tests. [97.301e]

T1A11 If you are a Technician licensee, what must you have to prove that you are authorized to use the Novice amateur frequencies below 30 MHz?
  A. A certificate from the FCC showing that you have notified them that you will be using the HF bands
  B. A certificate from an instructor showing that you have attended a class in HF communications
  C. Written proof of having passed a Morse code test
  D. No special proof is required before using the HF bands.

ANSWER C: When you pass your Morse Code test, you will be issued a certificate of successful completion (CSCE). Don't lose this! This is your proof that you have successfully passed the test. Keep it with your license. [97.301e]

T1B Emission privileges for Technician Class control operator, frequency selection and sharing, transmitter power

T1B01 At what point in your station is transceiver power measured?
A. At the power supply terminals inside the transmitter or amplifier
B. At the final amplifier input terminals inside the transmitter or amplifier
C. At the antenna terminals of the transmitter or amplifier
D. On the antenna itself, after the feed line

ANSWER C: Put your wattmeter into the antenna output of your transceiver. This is where you measure power output. [97.3b6]

T1B02 What is the term for the average power supplied to an antenna transmission line during one RF cycle at the crest of the modulation envelope?
A. Peak transmitter power
B. Peak output power
C. Average radio-frequency power
D. Peak envelope power

ANSWER D: If we measure to the crest of the modulation envelope, this is at its peak, and is called peak envelope power (PEP). [97.3b6]

T1B03 What is the maximum transmitting power permitted an amateur station in beacon operation?
A. 10 watts PEP output
B. 100 watts PEP output
C. 500 watts PEP output
D. 1500 watts PEP output

ANSWER B: Radio beacons are activated by amateur operators to automatically transmit a distinctive call sign on a specific beacon band frequency. This allows other hams to tune them in from far away to determine band conditions. If you can hear a distant beacon thousands of miles away, chances are you can communicate on that Amateur Radio band to other stations thousands of miles away in the direction of the propagational beacon. [97.203c]

Sequence (Minutes)	Call Sign	Station Location
00	4U1UN/B	United Nations, New York City
01	W6WX/B	Stanford University, CA
02	KH60/B	Honolulu Community College, HI
03	JA2IGY	JARL, Mt. Asama, Japan
04	4X6TU/B	Tel Aviv University, Israel
05	OH2B	Helsinki Technical University, Finland
06	CT3B	ARRM (Madeira Radio Society), Madeira Island
07	ZS6DN/B	Transvaal, South Africa
08	LU4AA/B	Radio Club Argentino, Buenos Aires
09	Silent	

Each station, in sequence, transmits for one minute on 14.01 MHz.
The 09 minute, the last in the sequence, is silent; then the sequence repeats.

Radio Beacon Stations

T1B04 If the FCC rules say that the amateur service is a secondary user of a frequency band, and another service is a primary user, what does this mean?
- A. Nothing special; all users of a frequency band have equal rights to operate.
- B. Amateurs are only allowed to use the frequency band during emergencies.
- C. Amateurs are allowed to use the frequency band only if they do not cause harmful interference to primary users.
- D. Amateurs must increase transmitter power to overcome any interference caused by primary users.

ANSWER C: We share the 900-MHz band with the vehicle locator service which is primary user of the frequencies. Same thing with 70 cm—we share it with military radio location services. They have first rights to these frequencies. [97.303]

T1B05 If you are using a frequency within a band assigned to the amateur service on a secondary basis, and a station assigned to the primary service on that band causes interference, what action should you take?
- A. Notify the FCC's regional Engineer in Charge of the interference.
- B. Increase your transmitter's power to overcome the interference.
- C. Attempt to contact the station and request that it stop the interference.
- D. Change frequencies; you may be causing harmful interference to the other station, in violation of FCC rules.

ANSWER D: Since our operation is on a secondary basis, we must change frequencies to keep from causing interference to the other station. [97.303]

T1B06 What rule applies if two amateur stations want to use the same frequency?
- A. The station operator with a lesser class of license must yield the frequency to a higher class licensee.
- B. The station operator with a lower power output must yield the frequency to the station with a higher power output.
- C. Both station operators have an equal right to operate on the frequency.
- D. Station operators in ITU Regions 1 and 3 must yield the frequency to stations in ITU Region 2.

ANSWER C: Hams must share the amateur frequencies. No ham owns a specific spot on the dial! All hams have an equal right to operate on any frequency that their class of license authorizes. [97.101b]

T1B07 What emission type may always be used for station identification, regardless of the transmitting frequency?
- A. CW
- B. RTTY
- C. MCW
- D. Phone

ANSWER A: CW stands for continuous wave. We use an interrupted continuous wave to transmit the dots and dashes of telegraphy. Telegraphy may be used for all station identification. [97.305a]

T1B08 On what frequencies within the 6-meter band may phone emissions be transmitted?
A. 50.0 - 54.0 MHz only
B. 50.1 - 54.0 MHz only
C. 51.0 - 54.0 MHz only
D. 52.0 - 54.0 MHz only

ANSWER B: In emission classification, "Phone" stands for voice modulation, and it's allowed throughout the 6-meter wavelength band except below 50.1 to 50.0 MHz. Always follow the band plan, page 15. [97.305c]

T1B09 On what frequencies within the 2-meter band may image emissions be transmitted?
A. 144.1 - 148.0 MHz only
B. 146.0 - 148.0 MHz only
C. 144.0 - 148.0 MHz only
D. 146.0 - 147.0 MHz only

ANSWER A: Image emissions (F3F) are allowed on the 2-meter wavelength band from 144.1 MHz to 148.0 MHz, and you have FM voice privileges throughout the band except below 144.1 MHz to 144.0 MHz. [97.305c]

T1B10 What is the maximum transmitting power permitted an amateur station on 146.52 MHz?
A. 200 watts PEP output
B. 500 watts ERP
C. 1000 watts DC input
D. 1500 watts PEP output

ANSWER D: When you upgrade to Technician, your new privileges will incorporate the popular 2-meter band. Would you believe you can run up to 1-1/2 kilowatts (1500 watts) on 2 meters? This is perfectly legal, but used only for those communications that need high power. This would include moon bounce, long-haul tropospheric ducting, Sporadic-E contacts, and meteor shower contacts. [97.313b]

A Professional Wattmeter

T1B11 Which band may NOT be used by Earth stations for satellite communications?
A. 6 meters
B. 2 meters
C. 70 centimeters
D. 23 centimeters

ANSWER A: There are no uplink frequencies on the 6-meter band for satellite communications. [97.209b2]

T1C Digital communications, station identification, ID with CSCE

T1C01 If you are a Novice licensee with a Certificate of Successful Completion of Examination (CSCE) for Technician privileges, how do you identify your station when transmitting on 146.34 MHz?
   A. You must give your call sign, followed by any suitable word that denotes the slant mark and the identifier "KT".
   B. You may not operate on 146.34 until your new license arrives.
   C. No special form of identification is needed.
   D. You must give your call sign and the location of the VE examination where you obtained the CSCE.

ANSWER A: You can go on the air immediately after you pass your 25 multiple-choice question examination. You must use a special designator after your Novice call sign to identify your recent upgrade. It takes about 60 days for your new license to arrive. You may elect to keep your same entry-level call sign, or trade it in for a slightly shorter Technician/General Class call. When your new call sign arrives, you won't need to identify with your temporary identifier code anymore. [97.119e1]

T1C02 What is the maximum frequency shift permitted for RTTY or data transmissions below 50 MHz?
   A. 0.1 kHz
   B. 0.5 kHz
   C. 1 kHz
   D. 5 kHz

ANSWER C: When you listen to RTTY, you will hear two distinct tones, mark and space. The maximum separation allowed is 1000 hertz, but 170 hertz is typical. [97.307f3/4]

T1C03 What is the maximum frequency shift permitted for RTTY or data transmissions above 50 MHz?
   A. 0.1 kHz or the sending speed, in bauds, whichever is greater
   B. 0.5 kHz or the sending speed, in bauds, whichever is greater
   C. 5 kHz or the sending speed, in bauds, whichever is greater
   D. The FCC rules do not specify a maximum frequency shift above 50 MHz.

ANSWER D: There are no rules for a specific maximum frequency shift above 50 MHz. [97.307]

T1C04 What is the maximum symbol rate permitted for packet transmissions on the 10-meter band?
   A. 300 bauds
   B. 1200 bauds
   C. 19.6 kilobauds
   D. 56 kilobauds

ANSWER B: You are going to love amateur radio and your home or portable computer system. They are a perfect match. Down on the 10-meter band, our maximum sending speed is 1200 bauds. [97.307f4]

T1C05 What is the maximum symbol rate permitted for packet transmissions on the 2-meter band?
A. 300 bauds
B. 1200 bauds
C. 19.6 kilobauds
D. 56 kilobauds
ANSWER C: The higher we go in frequency, the faster we may send our data transmissions. On 2 meters, we may step up to 19.6 kilobauds. [97.307f5]

T1C06 What is the maximum symbol rate permitted for RTTY or data transmissions between 28 and 50 MHz?
A. 56 kilobauds
B. 19.6 kilobauds
C. 1200 bauds
D. 300 bauds
ANSWER C: The lower the frequency, the lower the baud rate allowed. Don't be disappointed—1200 bauds is pretty quick. [97.307f4]

T1C07 What is the maximum symbol rate permitted for RTTY or data transmissions between 50 and 222 MHz?
A. 56 kilobauds
B. 19.6 kilobauds
C. 1200 bauds
D. 300 bauds
ANSWER B: The next possible answer up from 1200 bauds is 19.6 kilobauds (kilo means 1000). 19,600 bauds is real quick! [97.307f5]

T1C08 What is the maximum authorized bandwidth of RTTY, data or multiplexed emissions using an unspecified digital code within the frequency range of 50 to 222 MHz?
A. 20 kHz
B. 50 kHz
C. The total bandwidth shall not exceed that of a single-sideband phone emission.
D. The total bandwidth shall not exceed 10 times that of a CW emission.
ANSWER A: 20 kHz is the authorized bandwidth of a signal within this range. [97.307f5]

T1C09 What is the maximum symbol rate permitted for RTTY or data transmissions above 222 MHz?
A. 300 bauds
B. 1200 bauds
C. 19.6 kilobauds
D. 56 kilobauds
ANSWER D: 56 kilobauds (56,000) is about the fastest that one would expect to go above 222 MHz. [97.307f6]

T1C10 What is the maximum authorized bandwidth of RTTY, data or multiplexed emissions using an unspecified digital code within the frequency range of 222 to 450 MHz?

A. 50 kHz
B. 100 kHz
C. 150 kHz
D. 200 kHz

ANSWER B: 100 kHz is the maximum limit here. [97.307f6]

T1C11 What is the maximum authorized bandwidth of RTTY, data or multiplexed emissions using an unspecified digital code within the 70 cm amateur band?
A. 300 kHz
B. 200 kHz
C. 100 kHz
D. 50 kHz

ANSWER C: Between 420 and 450 MHz, bandwidth up to 100 kHz is permitted. [97.307f6]

T1D Correct language, phonetics, beacons and radio control of model craft and vehicles

T1D01 What is an amateur station called which transmits communications for the purpose of observation of propagation and reception?
A. A beacon
B. A repeater
C. An auxiliary station
D. A radio control station

ANSWER A: You can tune in radio beacons near 14.01 MHz, and on 10 meters between 28.2 MHz and 28.3 MHz. They use CW to send their call signs over and over again for propagation phenomena information. [97.3a9]

T1D02 What is the fastest code speed a repeater may use for automatic identification?
A. 13 words per minute
B. 20 words per minute
C. 25 words per minute
D. There is no limitation.

ANSWER B: 20 wpm is the exact speed for passing the Extra Class code test. Since hams are not required to know the code any faster than 20 wpm, that is the maximum rate for a repeater CW ID. [97.119b1]

T1D03 If you are using a language besides English to make a contact, what language must you use when identifying your station?
A. The language being used for the contact
B. The language being used for the contact, providing the US has a third-party communications agreement with that country
C. English
D. Any language of a country which is a member of the International Telecommunication Union

ANSWER C: It's perfectly legal to speak to another station in a foreign language; however, when it comes time to identify, use English. You can also identify in Morse code. [97.119b2]

T1D04 What do the FCC rules suggest you use as an aid for correct station identification when using phone?
A. A speech compressor
B. Q signals
C. A phonetic alphabet
D. Unique words of your choice

ANSWER C: When talking with a new station that is unfamiliar with your call sign, use the phonetic alphabet. This is especially helpful when communicating with a foreign station that may not speak good English. Everyone uses the same phonetics, and it makes it easy to pick out someone's call sign. [97.119b2]

Table T1D04. Phonetic Alphabet
Adopted by the International Telecommunication Union

A - Alpha	H - Hotel	O - Oscar	V - Victor
B - Bravo	I - India	P - Papa	W - Whiskey
C - Charlie	J - Juliette	Q - Quebec	X - X-Ray
D - Delta	K - Kilo	R - Romeo	Y - Yankee
E - Echo	L - Lima	S - Sierra	Z - Zulu
F - Foxtrot	M - Mike	T - Tango	
G - Golf	N - November	U - Uniform	

T1D05 What minimum class of amateur license must you hold to operate a beacon station?
A. Novice
B. Technician
C. General
D. Amateur Extra

ANSWER B: Everyone but the Novice may operate a beacon station. [97.203a]

T1D06 If a repeater is causing harmful interference to another repeater and a frequency coordinator has recommended the operation of one station only, who is responsible for resolving the interference?
A. The licensee of the unrecommended repeater
B. Both repeater licensees
C. The licensee of the recommended repeater
D. The frequency coordinator

ANSWER A: It's important to remember that all amateur repeaters must receive coordination. If you operate an amateur repeater that is uncoordinated, you are responsible for resolving interference to a coordinated repeater. [97.205c]

T1D07 If a repeater is causing harmful interference to another amateur repeater and a frequency coordinator has recommended the operation of both stations, who is responsible for resolving the interference?
A. The licensee of the repeater which has been recommended for the longest period of time
B. The licensee of the repeater which has been recommended the most recently
C. The frequency coordinator
D. Both repeater licensees

ANSWER D: If both repeater stations are coordinated, both repeater licensees must mutually work out the interference problem. [97.205c]

T1D08 If a repeater is causing harmful interference to another repeater and a frequency coordinator has NOT recommended either station, who is primarily responsible for resolving the interference?

A. Both repeater licensees
B. The licensee of the repeater which has been in operation for the longest period of time
C. The licensee of the repeater which has been in operation for the shortest period of time
D. The frequency coordinator

ANSWER A: In this case, both repeaters are uncoordinated, so both licensees must resolve the problem. [97.205c]

T1D09 What minimum information must be on a label affixed to a transmitter used for telecommand (control) of model craft?

A. Station call sign
B. Station call sign and the station licensee's name
C. Station call sign and the station licensee's name and address
D. Station call sign and the station licensee's class of license

ANSWER C: With your Technician Class license, you get to fly the coveted black flag. The black flag indicates 6-meter ham operation. Just make sure you have all of your license information on the side of your transmitter. [97.215a]

T1D10 What are the station identification requirements for an amateur transmitter used for telecommand (control) of model craft?

A. Once every ten minutes
B. Once every ten minutes, and at the beginning and end of each transmission
C. At the beginning and end of each transmission
D. Station identification is not required if the transmitter is labeled with the station licensee's name, address and call sign.

ANSWER D: Conceivably you could identify by letting your plane send your call sign in smoke signals and skywriting, but this is pretty crazy. No identification is required. [97.215a]

Old Channels			New Channels	
Freq. (MHz)	Channel I.D.		Freq. (MHz)	Channel I.D.
53.1	Black	Brown	50.80	00
53.2	Black	Red	50.82	01
53.3	Black	Orange	50.84	02
53.4	Black	Yellow	50.86	03
53.5	Black	Green	50.88	04
53.6	Black	Blue	50.90	05
53.7	Black	Violet	50.92	06
53.8	Black	Grey	50.94	07
53.9	Black	White	50.96	08
			50.98	09

Radio Control Channels

T1D11 What is the maximum transmitter power an amateur station is allowed when used for telecommand (control) of model craft?

A. One milliwatt
B. One watt
C. Two watts
D. Three watts

ANSWER B: If you ran more than one watt of power, every model in the country might take its command from your transmitter! One watt is a good power level to keep your model going within eyesight. [97.215c]

T1E Emergency communications; broadcasting; permissible one-way, satellite and third-party communication; indecent and profane language

T1E01 What is meant by the term broadcasting?
A. Transmissions intended for reception by the general public, either direct or relayed
B. Retransmission by automatic means of programs or signals from non-amateur stations
C. One-way radio communications, regardless of purpose or content
D. One-way or two-way radio communications between two or more stations

ANSWER A: You may not operate your station like an AM or shortwave broadcast station. You cannot transmit to the public directly. [97.3a10]

T1E02 Which of the following one-way communications may not be transmitted in the amateur service?
A. Telecommands to model craft
B. Broadcasts intended for the general public
C. Brief transmissions to make adjustments to the station
D. Morse code practice

ANSWER B: This is one of those questions where you are to look for the "not" answer. No, you may not broadcast information intended for the general public. Only commercial broadcast stations may do that. [97.3a10]

T1E03 What kind of payment is allowed for third-party messages sent by an amateur station?
A. Any amount agreed upon in advance
B. Donation of equipment repairs
C. Donation of amateur equipment
D. No payment of any kind is allowed.

ANSWER D: You may not receive payment for handling any type of third-party traffic. This includes payment for long-distance charges incurred during the third-party traffic. [97.113b]

T1E04 When may you send obscene words from your amateur station?
A. Only when they do not cause interference to other communications
B. Never; obscene words are prohibited in amateur transmissions.
C. Only when they are not retransmitted through a repeater
D. Any time, but there is an unwritten rule among amateurs that they should not be used on the air.

ANSWER B: No ham operator likes talking to a potty mouth. Don't use expletives over the airwaves. [97.113d]

T1E05 When may you send indecent words from your amateur station?
A. Only when they do not cause interference to other communications
B. Only when they are not retransmitted through a repeater

C. Any time, but there is an unwritten rule among amateurs that they should not be used on the air.

D. Never; indecent words are prohibited in amateur transmissions.

ANSWER D: Any type of indecent language is frowned upon by other hams. Remember, ham radio is a family hobby, and there may be young kids out there listening. [97.113d]

T1E06 When may you send profane words from your amateur station?

A. Only when they do not cause interference to other communications

B. Only when they are not retransmitted through a repeater

C. Never; profane words are prohibited in amateur transmissions.

D. Any time, but there is an unwritten rule among amateurs that they should not be used on the air.

ANSWER C: Using profane language is just not necessary on the ham bands. Watch your tongue, and communicate like a professional. [97.113d]

T1E07 If you wanted to use your amateur station to retransmit communications between a space shuttle and its associated Earth stations, what agency must first give its approval?

A. The FCC in Washington, DC

B. The office of your local FCC Engineer In Charge (EIC)

C. The National Aeronautics and Space Administration

D. The Department of Defense

ANSWER C: NASA usually gives everyone permission to retransmit Space Shuttle audio. When you hear the Space Shuttle transmissions, you are hearing them from a local repeater with permission to retransmit the NASA communications. [97.113e]

T1E08 When are third-party messages allowed to be sent to a foreign country?

A. When sent by agreement of both control operators

B. When the third party speaks to a relative

C. They are not allowed under any circumstances.

D. When the US has a third-party agreement with the foreign country or the third party is qualified to be a control operator

ANSWER D: Your ham radio station is not a substitute for the regular international telephone service. If a third party wishes to use your ham station to talk with another ham in a foreign country (with which there is a third-party agreement), the third party's communications must be of a personal nature and relatively unimportant. [97.115a2]

T1E09 If you let an unlicensed third party use your amateur station, what must you do at your station's control point?

A. You must continuously monitor and supervise the third party's participation.

B. You must monitor and supervise the communication only if contacts are made in countries which have no third-party communications agreement with the US.

C. You must monitor and supervise the communication only if contacts are made on frequencies below 30 MHz.

D. You must key the transmitter and make the station identification.

ANSWER A: Don't even leave the room when a third party is using your ham set. Your license is at stake, so stay right there with the third party to insure compliance with all FCC rules. See the Appendix for a list of third-party countries. [97.115b1]

T1E10 If a disaster disrupts normal communication systems in an area where the amateur service is regulated by the FCC, what kinds of transmissions may stations make?

    A. Those which are necessary to meet essential communication needs and facilitate relief actions

    B. Those which allow a commercial business to continue to operate in the affected area

    C. Those for which material compensation has been paid to the amateur operator for delivery into the affected area

    D. Those which are to be used for program production or newsgathering for broadcasting purposes

ANSWER A: If you do take part in emergency communications, keep your transmissions as short as possible. Listen to airline pilots communications over the airwaves—you should adopt their brief style when taking part in emergency communications. [97.401a]

T1E11 What information is included in an FCC declaration of a temporary state of communication emergency?

    A. A list of organizations authorized to use radio communications in the affected area

    B. A list of amateur frequency bands to be used in the affected area

    C. Any special conditions and special rules to be observed during the emergency

    D. An operating schedule for authorized amateur emergency stations

ANSWER C: If you are asked to stop transmitting on a certain frequency because it is reserved only for emergency communications, then by all means comply! Do listen in to see if there is anything that you might do to help—but avoid transmitting on the frequency unless directed to do so by the emergency net controller. If they are not asking for outside help, then don't transmit an offer for assistance. [97.401c]

Subelement T2 – Operating Procedures

3 exam questions
3 topic groups

T2A Repeater operation, courteous operation

T2A01 How do you call another station on a repeater if you know the station's call sign?

    A. Say "break, break 79," then say the station's call sign.

    B. Say the station's call sign, then identify your own station.

    C. Say "CQ" three times, then say the station's call sign.

    D. Wait for the station to call "CQ," then answer it.

ANSWER B: When we QSO with a station, we communicate with it. That's what QSO means. If the repeater is available, simply call the other station by its call sign, and then give your own call sign.

T2A02 Why should you pause briefly between transmissions when using a repeater?

    A. To check the SWR of the repeater

    B. To reach for pencil and paper for third-party communications

    C. To listen for anyone wanting to break in

    D. To dial up the repeater's autopatch

ANSWER C: A repeater is like a party line—there may be others who may wish to use the system. In an emergency, stations may break in saying "Break, Break, Break". Give up the channel immediately. Always leave enough time between picking up the conversation for other stations to break in. It's a pause that may refresh someone else's day in an emergency.

T2A03 Why should you keep transmissions short when using a repeater?

    A. A long transmission may prevent someone with an emergency from using the repeater.

    B. To see if the receiving station operator is still awake

    C. To give any listening non-hams a chance to respond

    D. To keep long distance charges down

ANSWER A: During peak traffic hours, keep your transmissions short. Repeaters are a great way to find out traffic reports and for reporting traffic accidents.

T2A04 What is the proper way to break into a conversation on a repeater?

    A. Wait for the end of a transmission and start calling the desired party.

    B. Shout, "break, break!" to show that you're eager to join the conversation.

    C. Turn on an amplifier and override whoever is talking.

    D. Say your call sign during a break between transmissions.

ANSWER D: A good way to join an ongoing repeater conversation is to quickly drop the last two letters of your call sign in between the transmissions. Remember, other stations will always leave a pause to allow stations to join in on the QSO.

T2A05 What is the purpose of repeater operation?

    A. To cut your power bill by using someone else's higher power system

    B. To help mobile and low-power stations extend their usable range

    C. To transmit signals for observing propagation and reception

    D. To make calls to stores more than 50 miles away

ANSWER B: Repeaters are sponsored by ham radio clubs and individual hams for everyone to use. They are usually placed high atop a mountain or a very tall building. Mobile and portable sets operate through repeaters with dramatically extended range. Even base stations are permitted to use repeaters for added communications distance. There is usually no charge for joining a repeater group. Some repeaters have autopatch, and those repeaters may require special access codes and financial support.

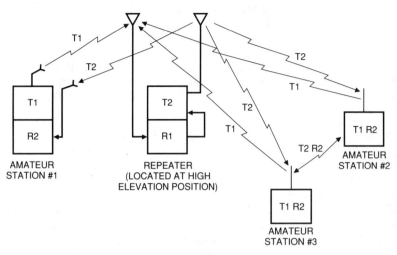

Repeater
Source: *Mobile 2-Way Radio Communications*, G.West, © 1993, Master Publishing, Inc.

T2A06 What causes a repeater to "time out"?
A. The repeater's battery supply runs out.
B. Someone's transmission goes on longer than the repeater allows.
C. The repeater gets too hot and stops transmitting until its circuitry cools off.
D. Something is wrong with the repeater.

ANSWER B: If you are long-winded, you could time out a repeater. Most repeaters have a one minute timer to limit any one transmission. If you out-talk the timer, the repeater shuts off, and won't come back on for about a minute or so. If you must really be long-winded, allow the repeater to reset by momentarily dropping your carrier until it goes "beep".

T2A07 During commuting rush hours, which type of repeater operation should be discouraged?
A. Mobile stations
B. Low-power stations
C. Highway traffic information nets
D. Third-party communications nets

ANSWER D: It's not a good idea to let a friend talk over your microphone as a third party during heavy repeater use time. During rush hours, most repeaters are used for traffic advisories and traffic accident reports.

T2A08 What is a courtesy tone (used in repeater operations)?
A. A sound used to identify the repeater
B. A sound used to indicate when a transmission is complete
C. A sound used to indicate that a message is waiting for someone
D. A sound used to activate a receiver in case of severe weather

ANSWER B: Most repeaters have a beep tone that lets you know when the other person has stopped transmitting. Wait at least a second before continuing to pick up the conversation.

T2A09 What is the meaning of: "Your signal is full quieting..."?
  A. Your signal is strong enough to overcome all receiver noise.
  B. Your signal has no spurious sounds.
  C. Your signal is not strong enough to be received.
  D. Your signal is being received, but no audio is being heard.

ANSWER A: When transmitting on VHF or UHF FM equipment, the S meter is simply a row of LCD bars that may illustrate relative signal strength. It's much easier to relate signal strength to how well the signal is quieting the background white noise.

T2A10 How should you give a signal report over a repeater?
  A. Say what your receiver's S meter reads.
  B. Always say: "Your signal report is five five..."
  C. Say the amount of signal quieting into the repeater.
  D. Try to imitate the sound quality you are receiving.

ANSWER C: The station with whom you are communicating wishes to know their signal strength into the repeater. They are not concerned with how well you hear the repeater, but rather how well you hear their station into the repeater. Signal quieting is the best method to illustrate signal strength.

T2A11 What is a repeater called which is available for anyone to use?
  A. An open repeater
  B. A closed repeater
  C. An autopatch repeater
  D. A private repeater

ANSWER A: Over 80 percent of United States 2-meter repeaters are open and available for anyone to use. These repeaters are listed in the American Radio Relay League (ARRL) repeater directory—a very handy book to have when traveling.

T2A12 What is the usual input/output frequency separation for repeaters in the 2-meter band?
  A. 600 kHz
  B. 1.0 MHz
  C. 1.6 MHz
  D. 5.0 MHz

ANSWER A: You will have this question, or one of the next three questions, on your exam, for sure. These are important to remember. Although some sets already come with the repeater splits memorized, some sets require initializing. On the 2-meter wavelength band, repeater inputs and outputs are usually separated by 600 kHz.

T2A13 What is the usual input/output frequency separation for repeaters in the 1.25-meter band?
  A. 600 kHz
  B. 1.0 MHz
  C. 1.6 MHz
  D. 5.0 MHz

ANSWER C: The 1.25-meter wavelength band is 222 MHz, and repeater separation is 1.6 MHz.

T2A14 What is the usual input/output frequency separation for repeaters in the 70-centimeter band?
A. 600 kHz
B. 1.0 MHz
C. 1.6 MHz
D. 5.0 MHz

ANSWER D: 70 centimeters (0.70 meters) is the 450-MHz band, and input and output repeater separation is 5 MHz.

T2A15 Why should local amateur communications use VHF and UHF frequencies instead of HF frequencies?
A. To minimize interference on HF bands capable of long distance communication
B. Because greater output power is permitted on VHF and UHF
C. Because HF transmissions are not propagated locally
D. Because signals are louder on VHF and UHF frequencies

ANSWER A: We use VHF and UHF frequencies for operating on the 6-meter, 2-meter, 222-MHz, 450-MHz, and 1270-MHz FM ham bands. These bands are so high in frequency that they are line-of-sight to repeaters. We would use worldwide frequencies below 30 MHz for longer skywave range. If we wish to talk locally, we go to local VHF and UHF band frequencies.

T2A16 How might you join a closed repeater system?
A. Contact the control operator and ask to join.
B. Use the repeater until told not to.
C. Use simplex on the repeater input until told not to.
D. Write the FCC and report the closed condition.

ANSWER A: Closed repeaters are available for membership. Most closed repeaters offer autopatch, paging, and sometimes remote high-frequency base functions. These exotic systems are supported by club membership. Listen to the repeater for announcements for their next general membership meeting. Come to the meeting, and they'll probably sign you up on the spot.

T2A17 How can on-the-air interference be minimized during a lengthy transmitter testing or loading up procedure?
A. Choose an unoccupied frequency.
B. Use a dummy load.
C. Use a non-resonant antenna.
D. Use a resonant antenna that requires no loading-up procedure.

ANSWER B: If you work on your transmitter a lot, you may wish to test it without actually putting a signal out on the airwaves. A dummy antenna might be a light bulb or a 50-ohm non-inductive resistance. The energy doesn't go very far. It's a great way to test your set before adding the antenna to it. It lets you run your rig at full output, but without actually going on the air. See question T4D03.

T2A18 What is the proper way to ask someone their location when using a repeater?
A. What is your QTH?
B. What is your 20?
C. Where are you?
D. Locations are not normally told by radio.

ANSWER C: Use plain language over repeater systems.

T2B Simplex operations, Q signals, RST signal reporting, repeater frequency coordination

T2B01 Why should simplex be used where possible, instead of using a repeater?

    A. Signal range will be increased.
    B. Long distance toll charges will be avoided.
    C. The repeater will not be tied up unnecessarily.
    D. Your antenna's effectiveness will be better tested.

ANSWER C: If you are close to the station you are communicating with, switch over to a direct channel. This is called simplex. Every band has several simplex channels specifically designed to relieve repeater congestion. You might be surprised how far simplex will go.

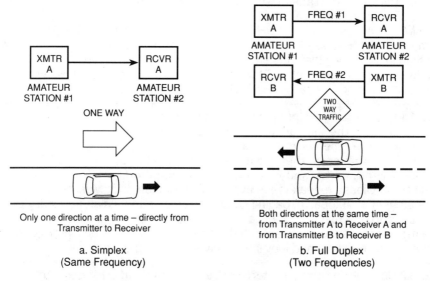

Only one direction at a time – directly from
Transmitter to Receiver

a. Simplex
(Same Frequency)

Both directions at the same time –
from Transmitter A to Receiver A and
from Transmitter B to Receiver B

b. Full Duplex
(Two Frequencies)

Simplex and Duplex Communications

T2B02 If you are talking to a station using a repeater, how would you find out if you could communicate using simplex instead?

    A. See if you can clearly receive the station on the repeater's input frequency.
    B. See if you can clearly receive the station on a lower frequency band.
    C. See if you can clearly receive a more distant repeater.
    D. See if a third station can clearly receive both of you.

ANSWER A: Almost all portable, mobile, and base VHF/UHF FM equipment have a small button marked "REV". Momentarily push the button, and it causes your receiver to quickly shift to the input frequency of the repeater. This allows you to see whether or not you can pick up the other station direct. If you can, suggest that you switch over to simplex for the remainder of the communications.

T2B03 If you are operating simplex on a repeater frequency, why would it be good amateur practice to change to another frequency?

A. The repeater's output power may ruin your station's receiver.
B. There are more repeater operators than simplex operators.
C. Changing the repeater's frequency is not practical.
D. Changing the repeater's frequency requires the authorization of the FCC.

ANSWER C: Double check that where you plan to operate simplex is a recognized simplex frequency. If you just pick any frequency, you could land right on a repeater input frequency. If you do, and if they ask you to move, please do so. Repeaters can't change frequency to avoid your communications.

T2B04 What is a repeater frequency coordinator?

A. Someone who organizes the assembly of a repeater station
B. Someone who provides advice on what kind of repeater to buy
C. The person whose call sign is used for a repeater's identification.
D. A person or group that recommends frequencies for repeater usage.

ANSWER D: Hams can put up their own repeaters, but only with coordination from a local repeater frequency coordinating committee. These committees insure that any new repeater system will fit in with existing systems.

T2B05 What is the proper Q signal to use to see if a frequency is in use before transmitting on CW?

A. QRV?
B. QRU?
C. QRL?
D. QRZ?

ANSWER C: See the Appendix for Q codes. For the Technician Class examination, you don't need to memorize all of the international Q signals, but you do need to know QRL. Just think of the letter "L" in QRL as in "Listening"—you listen before transmitting.

T2B06 What is one meaning of the Q signal "QSY"?

A. Change frequency.
B. Send more slowly.
C. Send faster.
D. Use more power.

ANSWER A: Think of the letter "Y" in QSY as a "Y" in the road. You must change direction—and in radio, QSY means change frequency.

T2B07 What is one meaning of the Q signal "QSO"?

A. A contact is confirmed.
B. A conversation is in progress.
C. A contact is ending.
D. A conversation is desired.

ANSWER B: A "QSO" is a communication between two hams. When there is a QSO going on, communications are in progress.

T2B08 What is the proper Q signal to use to ask if someone is calling you on CW?

A. QSL?
B. QRZ?

C. QRL?

D. QRT?

ANSWER B: If you are getting some zzzz's while listening to CW, you may miss the station's call sign that is calling you. QRZ means, "Who is calling?"

T2B09 What is the meaning of: "Your signal report is five seven..."?

A. Your signal is perfectly readable and moderately strong.

B. Your signal is perfectly readable, but weak.

C. Your signal is readable with considerable difficulty.

D. Your signal is perfectly readable with near pure tone.

ANSWER A: Looking at the RST signal report system chart, a 5 by 7 report is a pretty good indication that the other station is hearing you well. Since there is no third number, the report is for a voice transmission.

Table T2B09. RST Signal Reporting System

The RST System is a way of reporting on the quality of a received signal by using a three digit number. The first digit indicates *Readability* (R), the second digit indicates received *Signal Strength* (S) and the third digit indicates *Tone* (T).

READABILITY (R) Voice and CW
1 – Unreadable
2 – Barely readable, occasional words distinguishable
3 – Readable with considerable difficulty
4 – Readable with practically no difficulty
5 – Perfectly readable

SIGNAL STRENGTH (S) Voice and CW
1 – Faint and barely perceptible signals
2 – Very weak signals
3 – Weak signals
4 – Fair signals
5 – Fairly good signals
6 – Good signals
7 – Moderately strong signals
8 – Strong signals
9 – Extremely strong signals

*TONE (T) Use on CW only
1 – Very rough, broad signals, 60 cycle AC may be present
2 – Very rough AC tone, harsh, broad
3 – Rough, low pitched AC tone, no filtering
4 – Rather rough AC tone, some trace of filtering
5 – Filtered rectified AC note, musical, ripple modulated
6 – Slight trace of filtered tone but with ripple modulation
7 – Near DC tone but trace of ripple modulation
8 – Good DC tone, may have slight trace of modulation
9 – Purest, perfect DC tone with no trace of ripple or modulation.

* The TONE report refers only to the purity of the signal, and has no connection with its stability or freedom from clicks or chirps. If the signal has the characteristic steadiness of crystal control, add X to the report (e.g., RST 469X). If it has a chirp or "tail" (either on "make" or "break") add C (e.g., RST 469C). If it has clicks or other noticeable keying transients, add K (e.g., RST 469K). If a signal has both chirps and clicks, add both C and K (e.g., RST 469CK).

T2B10 What is the meaning of: "Your signal report is three three..."?

A. The contact is serial number thirty-three.

B. The station is located at latitude 33 degrees.

C. Your signal is readable with considerable difficulty and weak in strength.

D. Your signal is unreadable, very weak in strength

ANSWER C: Since only two numbers are given, you must be using voice. Better check out your equipment—a 3 by 3 report is not very good! See chart in question T2B09.

T2B11 What is the meaning of: "Your signal report is five nine plus 20 dB..."?

    A. Your signal strength has increased by a factor of 100.

    B. Repeat your transmission on a frequency 20 kHz higher.

    C. The bandwidth of your signal is 20 decibels above linearity.

    D. A relative signal-strength meter reading is 20 decibels greater than strength 9.

ANSWER D: Any signal over S9 is an excellent one. Most worldwide sets have well-calibrated S-meters that register 10, 20, 40, and 60 dB over S9. Your signal is plenty strong! Although the question and answer is worded technically correct, most hams would simply state that "your signal is 20 over 9". Same thing, but less formal on the air. See chart in question T2B09.

T2C Distress calling and emergency drills and communications - operations and equipment, Radio Amateur Civil Emergency Service (RACES)

T2C01 What is the proper distress call to use when operating phone?

    A. Say "MAYDAY" several times.

    B. Say "HELP" several times.

    C. Say "EMERGENCY" several times.

    D. Say "SOS" several times.

ANSWER A: The word "Mayday" deserves the highest priority. Always stand by and prepare to copy a "Mayday" message. See question T2C02.

T2C02 What is the proper distress call to use when operating CW?

    A. MAYDAY

    B. QRRR

    C. QRZ

    D. SOS

ANSWER D: When operating Morse Code CW, send "SOS" to indicate a grave emergency.

In a Blizzard, Ham Emergency Calls Have High Priority.

T2C03 What is the proper way to interrupt a repeater conversation to signal a distress call?

    A. Say "BREAK" twice, then your call sign.

    B. Say "HELP" as many times as it takes to get someone to answer.

    C. Say "SOS," then your call sign.

    D. Say "EMERGENCY" three times.

ANSWER A: On repeater frequencies, the word "break" is spoken several times to indicate a priority or emergency distress call. Keep this in mind when operating routinely on a repeater—don't say the word "break" unless it's an emergency or something very, very important.

T2C04 With what organization must you register before you can participate in RACES drills?
A. A local Amateur Radio club
B. A local racing organization
C. The responsible civil defense organization
D. The Federal Communications Commission

ANSWER C: RACES stands for Radio Amateur Civil Emergency Service. It is a division of the Civil Defense organization. You must be registered to take part in RACES drills.

T2C05 What is the maximum number of hours allowed per week for RACES drills?
A. One
B. Six, but not more than one hour per day
C. Eight
D. As many hours as you want

ANSWER A: No more than one hour may be allowed for RACES drills per week.

T2C06 How must you identify messages sent during a RACES drill?
A. As emergency messages
B. As amateur traffic
C. As official government messages
D. As drill or test messages

ANSWER D: To eliminate a misunderstanding of a drill message versus the real thing, always announce messages for practice as drill or test messages. You never know how many scanner monitor listeners are out there tuning in the ham bands!

T2C07 What is one reason for using tactical call signs such as "command post" or "weather center" during an emergency?
A. They keep the general public informed about what is going on.
B. They are more efficient and help coordinate public-service communications.
C. They are required by the FCC.
D. They increase goodwill between amateurs.

ANSWER B: It's perfectly legal to use such tactical words as "command post," "triage team," or "disaster communicator" during an emergency. This promotes efficiency in the ham radio communications being provided.

T2C08 What type of messages concerning a person's well-being are sent into or out of a disaster area?
A. Routine traffic
B. Tactical traffic
C. Formal message traffic
D. Health and Welfare traffic

ANSWER D: This type of traffic deserves priority because we are talking about the welfare of human lives.

T2C09 What are messages called which are sent into or out of a disaster area concerning the immediate safety of human life?
- A. Tactical traffic
- B. Emergency traffic
- C. Formal message traffic
- D. Health and Welfare traffic

ANSWER B: Any communications relating to the safety of human life or the immediate protection of property are considered emergency calls. They deserve the highest priority.

T2C10 Why is it a good idea to have a way to operate your amateur station without using commercial AC power lines?
- A. So you may use your station while mobile
- B. So you may provide communications in an emergency
- C. So you may operate in contests where AC power is not allowed
- D. So you will comply with the FCC rules

ANSWER B: Your author's station operates on solar panel power. The battery is safely outside, and the solar panels keep the battery charged even though he uses his ham station often.

T2C11 What is the most important accessory to have for a hand-held radio in an emergency?
- A. An extra antenna
- B. A portable amplifier
- C. Several sets of charged batteries
- D. A microphone headset for hands-free operation

ANSWER C: Rechargeable nickel cadmium batteries self-discharge up to 10 percent per week. This means a nickel cadmium battery will need frequent charging. Alkaline batteries have long shelf life; however, no recharge.

T2C12 Which type of antenna would be a good choice as part of a portable HF amateur station that could be set up in case of an emergency?
- A. A three-element quad
- B. A three-element Yagi
- C. A dipole
- D. A parabolic dish

ANSWER C: There is no simpler antenna than the common dipole. Remember the formula for constructing a dipole? Divide the frequency in MHz into 468 to obtain the end-to-end length in feet. Go back to your Novice questions and brush up on the equation used for the calculation.

Subelement T3 – Radio Wave Propagation

3 exam questions
3 topic groups

T3A Ionosphere, ionospheric regions, solar radiation

T3A01 What is the ionosphere?
- A. A area of the outer atmosphere where enough ions and free electrons exist to propagate radio waves
- B. A area between two air masses of different temperature and humidity, along which radio waves can travel

C. An ionized path in the atmosphere where lightning has struck

D. An area of the atmosphere where weather takes place

ANSWER A: The ionosphere is the electrified atmosphere from 40 miles to 400 miles above the earth. You can sometimes see it as "northern lights." It is charged up daily by the sun, and does some miraculous things to radio waves that strike it. Some radio waves are absorbed, during daylight hours, by the ionosphere's D layer. Others are bounced back to earth. Yet others penetrate the ionosphere, and never come back again. The wavelength of the radio waves determines whether the waves will be absorbed, refracted, or will penetrate. Here's a quick way to memorize what the different layers do during day and nighttime hours:

The D layer is about 40 miles up. The D layer is a Daylight layer; it almost disappears at night. D for Daylight. The D layer absorbs radio waves between 1 MHz to 7 MHz. These are long wavelengths. All others pass through.

The E layer is also a daylight layer, and it is very Eccentric. E for Eccentric. Patches of E layer ionization may cause some surprising reflections of signals on both high frequency as well as very high frequency. The E layer height is usually 70 miles.

The F1 layer is one of the layers furthest away. The F layer gives us those Far away signals. F for Far away. The F1 layer is present during daylight hours, and is up around 150 miles. The F2 layer is also present during daylight hours, and it gives us the Furthest range. The F2 layer is 250 miles high, and it's the best for the Furthest range on medium and short waves. At nighttime, the F1 and F2 layers combine to become just the F layer at 180 miles. This F layer at nighttime will usually bend radio waves between 1 MHz and 15 MHz back to earth. At night, the D and E layers disappear.

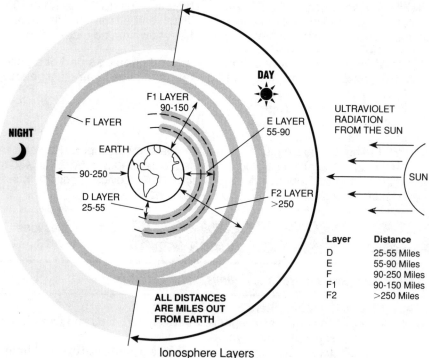

Layer	Distance
D	25-55 Miles
E	55-90 Miles
F	90-250 Miles
F1	90-150 Miles
F2	>250 Miles

ALL DISTANCES
ARE MILES OUT
FROM EARTH

Ionosphere Layers

Source: *Antennas – Selection and Installation,* © 1986, Master Publishing, Inc., Richardson, Texas

T3A02 What is the name of the area that makes long-distance radio communications possible by bending radio waves?

A. Troposphere
B. Stratosphere
C. Magnetosphere
D. Ionosphere

ANSWER D: When radio waves in the medium- and high-frequency range strike the ionosphere, they are often reflected, refracted, or simply bent back to earth. This occurs in the ionosphere.

T3A03 What causes the ionosphere to form?

A. Solar radiation ionizing the outer atmosphere
B. Temperature changes ionizing the outer atmosphere
C. Lightning ionizing the outer atmosphere
D. Release of fluorocarbons into the atmosphere

ANSWER A: The ionosphere surrounds us 24 hours a day. It is constantly "recharged" by the sun's solar radiation. When it gets "overcharged," you may see the effects as the Northern Lights.

T3A04 What type of solar radiation is most responsible for ionization in the outer atmosphere?

A. Thermal
B. Ionized particle
C. Ultraviolet
D. Microwave

ANSWER C: It's the ultraviolet component of the sun's radiation that creates our ionosphere. More ultraviolet radiation on any one day will lead us to either improved or disturbed radio conditions.

T3A05 Which ionospheric region limits daytime radio communications on the 80-meter band to short distances?

A. D region
B. E region
C. F1 region
D. F2 region

ANSWER A: The 80-meter wavelength band is around 3.75 MHz. This is absorbed by the D layer during daylight conditions. See question T3A01.

T3A06 Which ionospheric region is closest to the earth?

A. The A region
B. The D region
C. The E region
D. The F region

ANSWER B: Be careful on this question—they simply ask for the lowest ionospheric layer, not necessarily the lowest one that gives us skip. Since the D layer is the lowest, this is the correct answer. See question T3A01.

T3A07 Which ionospheric region most affects sky-wave propagation on the 6-meter band?

A. The D region
B. The E region

C. The F1 region

D. The F2 region

ANSWER B: A Technician Class no-code operator receives full privileges on the 6-meter band. During the summer months, Sporadic E skywave propagation allows for some exciting long-range contacts. See question T3A01.

T3A08 Which region of the ionosphere is the least useful for long-distance radio wave propagation?

A. The D region

B. The E region

C. The F1 region

D. The F2 region

ANSWER A: Just think of the D layer as that "Darn sponge" that absorbs long wavelength signals during daylight hours. Absorption always takes place in the D layer during daylight. See question T3A01.

T3A09 Which region of the ionosphere is mainly responsible for long-distance sky-wave radio communications?

A. D region

B. E region

C. F1 region

D. F2 region

ANSWER D: Since the F2 layer is higher than the F1 layer, it would be mainly responsible for the longest distance skywave hop back to earth. See question T3A01.

T3A10 What two sub-regions of ionosphere exist only in the daytime?

A. Troposphere and stratosphere

B. F1 and F2

C. Electrostatic and electromagnetic

D. D and E

ANSWER B: During daylight hours, the sun is so powerful it actually breaks the F layer into two distinct regions—F1 and F2. See question T3A01.

T3A11 Which two daytime ionospheric regions combine into one region at night?

A. E and F1

B. D and E

C. F1 and F2

D. E1 and E2

ANSWER C: Remember that F layer? It breaks apart into two distinct layers during the day, and recombines into one F layer at night. See question T3A01.

T3B Ionospheric absorption, causes and variation, maximum usable frequency

T3B01 Which region of the ionosphere is mainly responsible for absorbing radio signals during the daytime?

A. The F2 region

B. The F1 region

C. The E region

D. The D region

ANSWER D: The key word in this question is "absorption." Only one layer absorbs radio signals like a sponge, and that's the D layer during daylight hours. It absorbs medium- and low-frequency signals only. Higher frequencies pass through the D layer, and bounce off of other layers. See question T3A01.

T3B02 When does ionospheric absorption of radio signals occur?
A. When tropospheric ducting occurs
B. When long wavelength signals enter the D region
C. When signals travel to the F region
D. When a temperature inversion occurs

ANSWER B: We have most absorption to long wavelength signals. Long wavelength signals enter the D layer at low angles. Remember: "Lower-Longer." We'll use this memory jogger again for antenna length.

T3B03 What effect does the D region of the ionosphere have on lower-frequency HF signals in the daytime?
A. It absorbs the signals.
B. It bends the radio waves out into space.
C. It refracts the radio waves back to earth.
D. It has little or no effect on 80-meter radio waves.

ANSWER A: Here's that 80-meter wavelength band again during daylight hours—it's almost completely absorbed by the D layer.

T3B04 What causes the ionosphere to absorb radio waves?
A. The weather below the ionosphere
B. The ionization of the D region
C. The presence of ionized clouds in the E region
D. The splitting of the F region

ANSWER B: During daylight hours, the sun ionizes the D layer so that it absorbs long wavelength signals.

T3B05 What is the condition of the ionosphere just before local sunrise?
A. Atmospheric attenuation is at a maximum.
B. The D region is above the E region.
C. The E region is above the F region.
D. Ionization is at a minimum.

ANSWER D: The amount of ionization left in the ionosphere is at a minimum just before sunrise when the outside temperature is at its minimum.

T3B06 When is the ionosphere most ionized?
A. Dusk
B. Midnight
C. Midday
D. Dawn

ANSWER C: Maximum ionization occurs during midday when the daily temperature is usually maximum.

T3B07 When is the ionosphere least ionized?
A. Shortly before dawn
B. Just after noon
C. Just after dusk
D. Shortly before midnight

ANSWER A: Minimum ionization occurs just before the sun gets up. Maximum ionization occurs around midday.

T3B08 When is the E region most ionized?
A. Dawn
B. Midday
C. Dusk
D. Midnight

ANSWER B: Here's that ionization question again, and again! During midday, it's maximum; just before dawn, it's minimum.

T3B09 What happens to signals higher in frequency than the critical frequency?
A. They pass through the ionosphere.
B. They are absorbed by the ionosphere.
C. Their frequency is changed by the ionosphere to be below the maximum usable frequency.
D. They are reflected back to their source.

ANSWER A: The critical frequency rises during local daylight hours, and falls in the nighttime. If you transmit well above the critical frequency, your radio waves penetrate the ionosphere and don't refract back as a skywave.

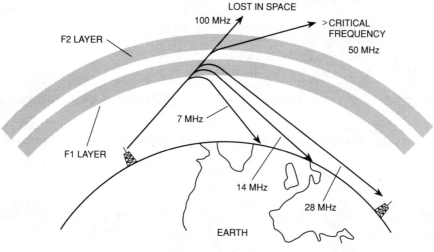

Critical Frequency

Source: *Antennas – Selection and Installation,* © 1986, Master Publishing, Inc., Richardson, Texas

T3B10 What causes the maximum usable frequency to vary?
A. The temperature of the ionosphere
B. The speed of the winds in the upper atmosphere
C. The amount of radiation received from the sun, mainly ultraviolet
D. The type of weather just below the ionosphere

ANSWER C: Worldwide band conditions will vary from day to day. Some changes, such as seasonal changes and time of day changes, are predictable. Others may be sudden due to intense solar activity on the sun creating changes in ultraviolet radiation on the earth. The sun's activity is important to monitor.

T3B11 What does maximum usable frequency mean?
  A. The highest frequency signal that will reach its intended destination
  B. The lowest frequency signal that will reach its intended destination
  C. The highest frequency signal that is most absorbed by the iono-
    sphere
  D. The lowest frequency signal that is most absorbed by the ionosphere
ANSWER A: Maximum usable frequency is abbreviated "MUF." Monthly maga-
zines predict the maximum usable frequency from here to there, anywhere in
the world, at any time.

T3C Propagation, including ionospheric, tropospheric, line-of-sight scatter propagation, and maximum usable frequency

**T3C01 What kind of propagation would best be used by two stations
within each other's skip zone on a certain frequency?**
  A. Ground-wave
  B. Sky-wave
  C. Scatter-mode
  D. Ducting
ANSWER C: It's quite possible your skywaves will actually skip over the station
you wish to communicate with. You might switch to a lower band to shorten
your range. Another trick is to turn on the power amplifier, and operate scatter
mode communications. The extra amount of transmit power will bounce signals
back to the normally vacant skip zone. Sometimes they bounce off to the side,
and sometimes they bounce ahead. All of this is known as back scatter, side
scatter, and forward scatter. If you have enough power and a directional
antenna, scatter communications will help fill in the skip zone.

**T3C02 If you are receiving a weak and distorted signal from a distant
station on a frequency close to the maximum usable frequency, what type
of propagation is probably occurring?**
  A. Ducting
  B. Line-of-sight
  C. Scatter
  D. Ground-wave
ANSWER C: It's unbelievable, but sometimes you can talk to stations in South
America with your antenna pointed west, and with better results in the evening
hours.

**T3C03 How are VHF signals propagated within the range of the visible
horizon?**
  A. By sky wave
  B. By direct wave
  C. By plane wave
  D. By geometric wave
ANSWER B: If you can see it, you can work it on VHF and UHF via direct wave.

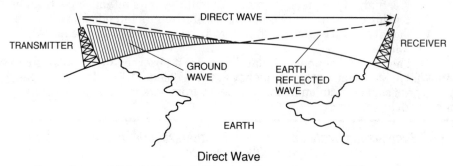

Direct Wave

Source: *Antennas – Selection and Installation,* © 1986, Master Publishing, Inc., Richardson, Texas

T3C04 Ducting occurs in which region of the atmosphere?
A. F2
B. Ectosphere
C. Troposphere
D. Stratosphere

ANSWER C: The key word in this question is "ducting." In your home you use ducts to shuffle that warm or cool air around the house. Out in radio land, natural atmospheric ducts form that shuffle VHF and UHF radio waves well beyond line-of-sight range. Tropospheric ducting occurs most often during the summer months, and sometimes occurs in the presence of large storm systems. Your author, Gordon West, is one of the record holders in tropospheric ducting on VHF line-of-sight frequencies between his home near Los Angeles all the way over to Hawaii. This is not skip off the ionosphere, but rather tropospheric ducting several hundred feet above the water all those thousands of miles away!

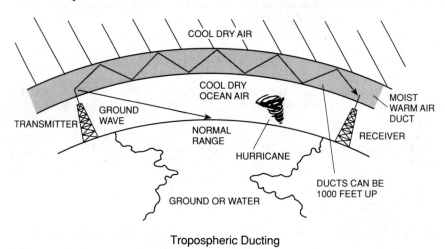

Tropospheric Ducting

T3C05 What effect does tropospheric bending have on 2-meter radio waves?
A. It lets you contact stations farther away.
B. It causes them to travel shorter distances.

C. It garbles the signal.

D. It reverses the sideband of the signal.

ANSWER A: The nice thing about tropospheric bending and ducting is it gives us some extraordinary range that we normally don't get under average weather conditions.

T3C06 What causes tropospheric ducting of radio waves?

A. A very low pressure area

B. An aurora to the north

C. Lightning between the transmitting and receiving stations

D. A temperature inversion

ANSWER D: Have you ever seen a mirage? Out on the desert, it looks like blue water instead of sand ahead. Actually that's the blue sky you are looking at. Light waves that normally travel in straight lines bounce off the super-heated windless sand and pavement and are reflected back to your eyes. Same thing during a tropospheric duct—but just backwards. Typically straight-line VHF and UHF signals begin to travel up and away, but are bent back by a sharp boundary layer of warm, moist air overlying cool, dry air below and above. In the city, this is what traps the smog and gives us one of those unbearable days. Get on the radio—it will be unbelievable. See figure at question T3C04.

T3C07 What causes VHF radio waves to be propagated several hundred miles over oceans?

A. A polar air mass

B. A widespread temperature inversion

C. An overcast of cirriform clouds

D. A high-pressure zone

ANSWER B: Some of these tropospheric ducts may extend out to 1000 miles. The record is between Gordon West's (your author) home in California and Hawaii. Any hams in the area are invited to stop by and see his record-breaking tropo station.

T3C08 In what frequency range does tropospheric ducting most often occur?

A. SW

B. MF

C. HF

D. VHF

ANSWER D: The 2-meter wavelength band is one of the best for taking advantage of tropospheric ducting conditions. With your Technician Class license around the corner, start planning for your 2-meter station now.

T3C09 In what frequency range does sky-wave propagation least often occur?

A. LF

B. MF

C. HF

D. VHF

ANSWER D: You will seldom find any skywave propagation on VHF frequencies above 144 MHz. It happens, but very infrequently.

T3C10 What weather condition may cause tropospheric ducting?

A. A stable high-pressure system
B. An unstable low-pressure system
C. A series of low-pressure waves
D. Periods of heavy rainfall

ANSWER A: You can watch your local weather maps for signs of temperature inversions. If they predict bad air quality, you can predict good band qualities on VHF and UHF.

T3C11 What band conditions might indicate long-range skip on the 6-meter and 2-meter bands?

A. Noise on the 80-meter band
B. The absence of signals on the 10-meter band
C. Very long-range skip on the 10-meter band
D. Strong signals on the 10-meter band from stations about 500-600 miles away

ANSWER D: A good indication that Sporadic E skywaves may be coming in on 6 meters, and at times on 2 meters, is when high-frequency signals on the 10-meter band pull in as close as 500 to 600 miles away.

Subelement T4 – Amateur Radio Practices

4 exam questions
4 topic groups

T4A Electrical wiring, including switch location, dangerous voltages and currents

T4A01 Where should the green wire in a three-wire AC line cord be connected in a power supply?

A. To the fuse
B. To the "hot" side of the power switch
C. To the chassis
D. To the white wire

ANSWER C: Green is ground. That's easy to remember. The chassis of our radio equipment is the metal top, bottom, and side covers that keep everything together on the inside. Most radio sets offer a ground connection on their rear. A green ground connection conductor will let you keep your colors straight.

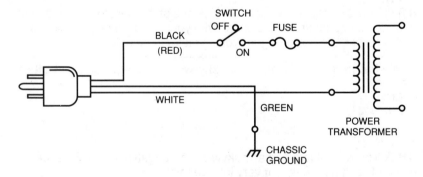

AC Line Connections

T4A02 Where should the black (or red) wire in a three-wire AC line cord be connected in a power supply?

    A. To the white wire

    B. To the green wire

    C. To the chassis

    D. To the fuse

ANSWER D: We always fuse the hot lead. Most ac line cords use a black— sometimes red—wire in the three-wire cord that plugs into the wall socket. This goes to the fuse. It's hot, so don't touch it when everything's plugged in. See figure at question T4A01.

T4A03 Where should the white wire in a three-wire AC line cord be connected in a power supply?

    A. To the side of the power transformer's primary winding that has a fuse

    B. To the side of the power transformer's primary winding that does not have a fuse

    C. To the chassis

    D. To the black wire

ANSWER B: The white wire is the other side of your hot wire line cord. Don't touch it, either, even though we will consider it "neutral." It goes to one side of the transformer that kicks up the voltage inside your set. See figure at question T4A01. Remember, *White* for transformer *Winding Without* a fuse.

T4A04 What document is used by almost every US city as the basis for electrical safety requirements for power wiring and antennas?

    A. The Code of Federal Regulations

    B. The Proceedings of the IEEE

    C. The ITU Radio Regulations

    D. The National Electrical Code

ANSWER D: It is the National Electrical Code that details proper electrical safety wiring requirements.

T4A05 What document would you use to see if you comply with standard electrical safety rules when building an amateur antenna?

    A. The Code of Federal Regulations

    B. The Proceedings of the IEEE

    C. The National Electrical Code

    D. The ITU Radio Regulations

ANSWER C: If you are planning on doing any electrical or antenna wiring, better check with the National Electrical Code to make sure you comply with all electrical safety standards.

T4A06 Where should fuses be connected on a mobile transceiver's DC power cable?

    A. Between the red and black wires

    B. In series with just the black wire

    C. In series with just the red wire

    D. In series with both the red and black wires

ANSWER D: We fuse both the red and the black wires to insure no ground currents are accidentally pulled through the black negative lead on your mobile installation. The fuse on the red wire prevents damage that would be caused by short circuits.

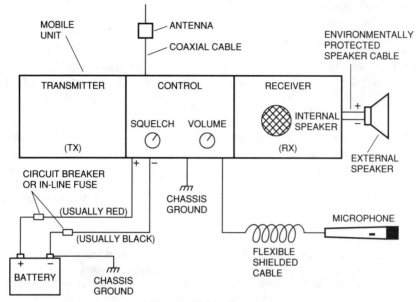

A Typical Mobile Unit Setup

Source: *Mobile 2-Way Radio Communications,* G.West, © 1993, Master Publishing, Inc.

T4A07 Why is the retaining screw in one terminal of a wall outlet made of brass while the other one is silver colored?
A. To prevent corrosion
B. To indicate correct wiring polarity
C. To better conduct current
D. To reduce skin effect

ANSWER B: The silver thread is the neutral connection, and the brass button is the hot terminal inside your light socket. We use different metals to keep track of polarity.

T4A08 How much electrical current flowing through the human body is usually fatal?
A. As little as 1/10 of an ampere
B. Approximately 10 amperes
C. More than 20 amperes
D. Current flow through the human body is never fatal.

ANSWER A: One-tenth of an ampere (amp) is the same as 100 milliamperes. One-tenth of an amp is as little as the small current drawn by a tiny dial light. It is enough to zap you for good if it travels through your body and heart in the right path. Never, never, never work without shoes on a concrete garage floor. Never let any metal electrical appliance get near a bathtub.

T4A09 Which body organ can be fatally affected by a very small amount of electrical current?
A. The heart
B. The brain

C. The liver

D. The lungs

ANSWER A: Your heart is a pump that is powered by your own electricity. If you disturb the natural flow of electricity driving the heart, it could be fatal. This is why you never work on electrical equipment in the garage with a cement floor when you are not wearing shoes. Electricity could flow through your hands, through your heart, and out of your feet to ground. Not good.

T4A10 How much electrical current flowing through the human body is usually painful?

A. As little as 1/500 of an ampere

B. Approximately 10 amperes

C. More than 20 amperes

D. Current flow through the human body is never painful.

ANSWER A: We usually begin to feel the sensation of pain with as little as 2 milliamperes of current. This is less current than some electric clocks require. Be extremely careful never to let any current pass through your body. When it gets to the heart, it may be all over for you.

T4A11 What is the minimum voltage which is usually dangerous to humans?

A. 30 volts

B. 100 volts

C. 1000 volts

D. 2000 volts

ANSWER A: Even a couple of golf cart batteries could kill you if you aren't careful. This is why you must be especially careful when leaning across a bank of batteries and allowing current to flow through your body accidentally.

T4A12 Where should the main power switch for a high-voltage power supply be located?

A. Inside the cabinet, to kill the power if the cabinet is opened

B. On the back side of the cabinet, out of sight

C. Anywhere that can be seen and reached easily

D. A high voltage power supply should not be switch-operated.

ANSWER C: Does everyone know where the main power switch is for your home? Let your whole family try turning off the power so they are familiar with the power line switch. If you have a special power switch in your ham shack, let everybody try their hand at turning it on and off.

T4A13 What precaution should you take when leaning over a power amplifier?

A. Take your shoes off.

B. Watch out for loose jewelry contacting high voltage.

C. Shield your face from the heat produced by the power supply.

D. Watch out for sharp edges which may snag your clothing.

ANSWER B: Gold and silver jewelry are highly conductive. Make sure that they do not contact high voltage.

T4A14 What is an important safety rule concerning the main electrical box in your home?

A. Make sure the door cannot be opened easily.

B. Make sure something is placed in front of the door so no one will be able to get to it easily.

C. Make sure others in your home know where it is and how to shut off the electricity.

D. Warn others in your home never to touch the switches, even in an emergency.

ANSWER C: At least once a year, run a practice exercise where everyone goes to your electrical panel and shuts off all the circuit breakers. Tell everyone to do this in case of a house fire or an electrocution. After the test, spend a couple of hours reprogramming your VCR and everything else that is blinking "12:00."

T4A15 What should you do if you discover someone who is being burned by high voltage?

A. Run from the area so you won't be burned too.

B. Turn off the power, call for emergency help and give CPR if needed.

C. Immediately drag the person away from the high voltage.

D. Wait for a few minutes to see if the person can get away from the high voltage on their own, then try to help.

ANSWER B: Does everyone know CPR? Run some practice exercises where you have a prescribed plan in case of an electrical accident at your QTH.

T4B Meters, including volt, amp, multi, peak-reading, RF watt and placement, and ratings of fuses and switches

T4B01 How is a voltmeter usually connected to a circuit under test?

A. In series with the circuit

B. In parallel with the circuit

C. In quadrature with the circuit

D. In phase with the circuit

ANSWER B: We test for voltage by hooking our meter *across* the voltage source without undoing any wires. This is called a parallel connection. Checking the voltage when operating equipment is called "checking the voltage source under load."

T4B02 How can the range of a voltmeter be increased?

A. By adding resistance in series with the circuit under test

B. By adding resistance in parallel with the circuit under test

C. By adding resistance in series with the meter, between the meter and the circuit under test

D. By adding resistance in parallel with the meter, between the meter and the circuit under test

ANSWER C: Let's say you want to test your new 12-volt power supply, but all you have is an old 6-volt panel meter. Through calculations (not required for this exam) you could add resistance in series with the meter to get 12 volts to register exactly 6 volts on your panel meter.

T4B03 What happens inside a voltmeter when you switch it from a lower to a higher voltage range?

A. Resistance is added in series with the meter.

B. Resistance is added in parallel with the meter.

C. Resistance is reduced in series with the meter.

D. Resistance is reduced in parallel with the meter.

ANSWER A: As more resistance is added in series with a voltmeter, the meter can indicate a higher voltage range without exceeding the meter's maximum rating. This is why you must look carefully on a multimeter to see to which voltage range the range switch is set. Many digital voltmeters have an auto-ranging circuit so you don't have to set the range.

T4B04 How is an ammeter usually connected to a circuit under test?

A. In series with the circuit

B. In parallel with the circuit

C. In quadrature with the circuit

D. In phase with the circuit

ANSWER A: An ammeter measures current. To measure current, turn off the power, disconnect one lead of the load from its source voltage and insert an ammeter in series with that lead. If you are measuring dc current, you will need to connect the meter with the correct polarity, so the meter reads up scale when power is turned on.

T4B05 How can the range of an ammeter be increased?

A. By adding resistance in series with the circuit under test

B. By adding resistance in parallel with the circuit under test

C. By adding resistance in series with the meter

D. By adding resistance in parallel with the meter

ANSWER D: An ammeter measures the flow of electrons through its terminals. The keyword here is "through" so we must remove a lead to the load to insert the ammeter into the current path. However, we have to bypass some of the current if it exceeds the basic rating of the ammeter. To bypass current, we add a precise, calibrated amount of resistance, called a shunt, in parallel with the meter. This allows the meter to indicate higher amounts of current accurately. *Remember this: For the voltmeter, add resistance in series to extend its range. For the ammeter, add resistance in parallel to extend its range.*

T4B06 What does a multimeter measure?

A. SWR and power

B. Resistance, capacitance and inductance

C. Resistance and reactance

D. Voltage, current and resistance

ANSWER D: Every amateur operator should own a multimeter. The multiple function meter can measure voltage, current, and resistance and check continuity. Even an inexpensive multimeter is better than no meter when you are trying to check out a circuit in the field. You can buy an excellent multimeter for less than $25.00 at your local Radio Shack store.

T4B07 Where should an RF wattmeter be connected for the most accurate readings of transmitter output power?

A. At the transmitter output connector

B. At the antenna feed point

C. One-half wavelength from the transmitter output

D. One-half wavelength from the antenna feed point

ANSWER A: Right at the transmitter. It's that simple.

T4B08 At what line impedance do most RF wattmeters usually operate?
   A. 25 ohms
   B. 50 ohms
   C. 100 ohms
   D. 300 ohms

ANSWER B: Since Amateur Radio transceivers all operate at a 50-ohm imped-ance, our wattmeter should also be rated at 50 ohms. The impedance is kept the same for a perfect match.

T4B09 What does a directional wattmeter measure?
   A. Forward and reflected power
   B. The directional pattern of an antenna
   C. The energy used by a transmitter
   D. Thermal heating in a load resistor

ANSWER A: A directional wattmeter is a special wattmeter that can simulta-neously measure forward and reflected power levels. Some directional wattmeters use only one needle, and a control that allows you to check for forward, and then reverse, power levels. They are a handy thing to have if you experiment with antennas.

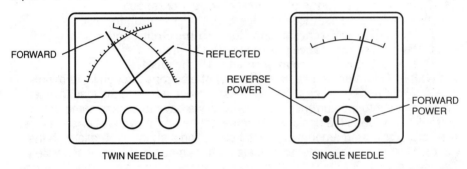

FORWARD   REFLECTED

REVERSE POWER

FORWARD POWER

TWIN NEEDLE   SINGLE NEEDLE

Directional Wattmeter

T4B10 If a directional RF wattmeter reads 90 watts forward power and 10 watts reflected power, what is the actual transmitter output power?
   A. 10 watts
   B. 80 watts
   C. 90 watts
   D. 100 watts

ANSWER B: You can buy an inexpensive twin-needle wattmeter that allows you to simultaneously monitor forward power and reverse power in watts. Any watts reflected back to the transmitter by a poor antenna are considered lost. They go up in transmitter heat or transmission line loss. See question T4B09.

Forward Power − Reflected Power = Actual Power Output
     90 W   −   10 W   =   80 W

T4B11 If a directional RF wattmeter reads 96 watts forward power and 4 watts reflected power, what is the actual transmitter output power?
   A. 80 watts
   B. 88 watts

C. 92 watts

D. 100 watts

ANSWER C: Simply subtract the wasted reflected power from the forward output power, and you can see what's actually reaching the antenna feedpoint. See question T4B09.

Forward Power − Reflected Power = Actual Power Output

96 W − 4 W = 92 W

T4C Marker generator, crystal calibrator, signal generators and impedance-match indicator

T4C01 What is a marker generator?

A. A high-stability oscillator that generates reference signals at exact frequency intervals

B. A low-stability oscillator that "sweeps" through a range of frequencies

C. A low-stability oscillator used to inject a signal into a circuit under test

D. A high-stability oscillator which can produce a wide range of frequencies and amplitudes

ANSWER A: Most worldwide transceivers have a built-in marker generator that, when turned on, sends out a tone at 100-kHz intervals. This can be useful in calibrating your equipment. It's also useful for the visually impaired. We use these tones as a reference signal.

T4C02 How is a marker generator used?

A. To calibrate the tuning dial on a receiver

B. To calibrate the volume control on a receiver

C. To test the amplitude linearity of a transmitter

D. To test the frequency deviation of a transmitter

ANSWER A: Here's that marker generator again—this time they want to know how it is used. Use it to mark the band edges when operating mobile. If you use it correctly, it allows you to calibrate the tuning dial all in your head without looking at the dial.

T4C03 What device is used to inject a frequency calibration signal into a receiver?

A. A calibrated voltmeter

B. A calibrated oscilloscope

C. A calibrated wavemeter

D. A crystal calibrator

ANSWER D: On old tuning dial receivers, the dial cord would sometimes break. After replacing the dial cord, a crystal calibrator was used to synchronize the tuning dial with the actual frequency. You're not likely to find a dial cord in modern sets.

T4C04 What frequency standard may be used to calibrate the tuning dial of a receiver?

A. A calibrated voltmeter

B. Signals from WWV and WWVH

C. A deviation meter

D. A sweep generator

ANSWER B: WWV signals from Boulder, Colorado, are easily picked up at 2.5 MHz, 5 MHz, 10 MHz, 15 MHz, and 20 MHz. WWVH transmits from Hawaii. They are on the same frequencies, so you should have no problem picking up either one of these stations.

T4C05 How might you check the accuracy of your receiver's tuning dial?

A. Tune to the frequency of a shortwave broadcasting station.

B. Tune to a popular amateur net frequency.

C. Tune to one of the frequencies of station WWV or WWVH.

D. Tune to another amateur station and ask what frequency the operator is using.

ANSWER C: There is nothing more accurate than WWV signals. That's why they are there, on the air, 24 hours a day. You can easily hear their distinctive once-a-second tick, and their every-one-minute time broadcast. Also, listen at 18 minutes past the hour for solar conditions.

T4C06 What device produces a stable, low-level signal that can be set to a desired frequency?

A. A wavemeter

B. A reflectometer

C. A signal generator

D. An oscilloscope

ANSWER C: A signal generator with a digital readout is useful in troubleshooting scanner radios. You can inject a tiny signal into the "front end" of the set to check it's operation and to be sure it is on frequency.

T4C07 What is an RF signal generator used for?

A. Measuring RF signal amplitudes

B. Aligning tuned circuits

C. Adjusting transmitter impedance-matching networks

D. Measuring transmission line impedances

ANSWER B: Your author uses an RF signal generator in his shack for aligning scanners, shortwave sets, and hand-held portables. They have tuned circuits. A tuned circuit is one that is designed to select signals of certain frequencies, and reject others.

T4C08 What device can measure an impedance mismatch in your antenna system?

A. A field-strength meter

B. An ammeter

C. A wavemeter

D. A reflectometer

ANSWER D: Impedance is a word that describes the opposition to the flow of electrons in an alternating current (ac) circuit. Most circuits have a specific impedance, and most Amateur Radio antennas should have an impedance of 50 ohms. If your antenna is perfectly matched to your coax and transceiver, everything will be a 50-ohm impedance match, and you will have minimum reflected power. You can verify this with an SWR reflectometer.

T4C09 Where should a reflectometer be connected for best accuracy when reading the impedance match between an antenna and its feed line?

A. At the antenna feed point
B. At the transmitter output connector
C. At the midpoint of the feed line
D. Anywhere along the feed line

ANSWER A: An SWR meter is a reflectometer. If you do a lot of experimenting with antennas, and we hope you will, buy an external SWR meter. Most worldwide radios have a built-in SWR meter that measures the SWR at the transmitter after it travels all the way back down the feed line. Take a portable SWR analyzer to the antenna feed point, and measure SWR truly presented by the antenna itself.

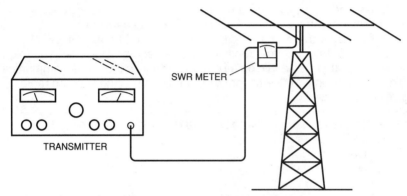

Measure SWR Right at Antenna

T4C10 If you use a 3-30 MHz RF power meter for VHF, how accurate will its readings be?

A. They will not be accurate.
B. They will be accurate enough to get by.
C. If it properly calibrates to full scale in the set position, they may be accurate.
D. They will be accurate providing the readings are multiplied by 4.5.

ANSWER A: You cannot reliably use a CB-radio-type power meter for 2-meter VHF. You must use a VHF/UHF power meter for best results.

T4C11 If you use a 3-30 MHz SWR meter for VHF, how accurate will its readings be?

A. They will not be accurate.
B. They will be accurate enough to get by.
C. If it properly calibrates to full scale in the set position, they may be accurate.
D. They will be accurate providing the readings are multiplied by 4.5.

ANSWER C: Most CB-type SWR bridges *sometimes* will work on 2-meter VHF frequencies. You should buy a special meter for these higher frequencies.

T4D Dummy antennas, S-meter, exposure of the human body to RF

T4D01 What device should be connected to a transmitter's output when you are making transmitter adjustments?
A. A multimeter
B. A reflectometer
C. A receiver
D. A dummy antenna

ANSWER D: A dummy antenna is used when you want to load your transmitter at full power output, but not transmit the signal very far.

T4D02 What is a dummy antenna?
A. An nondirectional transmitting antenna
B. A nonradiating load for a transmitter
C. An antenna used as a reference for gain measurements
D. A flexible antenna usually used on hand-held transceivers

ANSWER B: A dummy antenna does not radiate very far. However, you might be surprised how far you can talk on a light bulb. Your author once communicated 3000 miles away on a 100-watt light bulb, much to his and his ham radio students' amazement.

T4D03 What is the main component of a dummy antenna?
A. A wire-wound resistor
B. An iron-core coil
C. A noninductive resistor
D. An air-core coil

ANSWER C: You can buy a dummy load that will handle up to 100 watts for about $20. You can make one yourself if you find some high-wattage resistors that don't use a coil of wire as their resistance element. Light bulbs are handy because you can observe the results of your adjustments by the intensity of the light; however, light bulbs tend to act like an inductor (a coil of wire), so they are not the best. Any non-inductive resistor combination will do the job nicely.

T4D04 What device is used in place of an antenna during transmitter tests so that no signal is radiated?
A. An antenna matcher
B. A dummy antenna
C. A low-pass filter
D. A decoupling resistor

ANSWER B: Here's yet another question about that antenna that doesn't radiate—the dummy load or dummy antenna.

T4D05 Why would you use a dummy antenna?
A. For off-the-air transmitter testing
B. To reduce output power
C. To give comparative signal reports
D. To allow antenna tuning without causing interference

ANSWER A: Another question about the dummy load—just remember, it doesn't radiate, and allows you to run your transmitter full bore without causing interference.

T4D06 What minimum rating should a dummy antenna have for use with a 100-watt single-sideband phone transmitter?
A. 100 watts continuous
B. 141 watts continuous
C. 175 watts continuous
D. 200 watts continuous
ANSWER A: Your single-sideband phone transmitter will put out about 100 watts. This means you need a dummy load capable of handling that power—100 watts continuous.

T4D07 Why might a dummy antenna get warm when in use?
A. Because it stores electric current
B. Because it stores radio waves
C. Because it absorbs static electricity
D. Because it changes RF energy into heat
ANSWER D: It's perfectly normal for a dummy load to begin to get warm after a few seconds of transmitting. Some dummy loads also have a red warning light to indicate overheating. If the light illuminates, stop transmitting. Let the dummy load cool before you begin your testing again.

T4D08 What is used to measure relative signal strength in a receiver?
A. An S meter
B. An RST meter
C. A signal deviation meter
D. An SSB meter
ANSWER A: An S meter measures signal strength. All worldwide ham sets have one.

T4D09 How can exposure to a large amount of RF energy affect body tissue?
A. It causes radiation poisoning.
B. It heats the tissue.
C. It paralyzes the tissue.
D. It produces genetic changes in the tissue.
ANSWER B: It's the heating effect on body tissue that causes permanent damage.

T4D10 Which body organ is the most likely to be damaged from the heating effects of RF radiation?
A. Eyes
B. Hands
C. Heart
D. Liver
ANSWER A: The eyes are most sensitive to RF radiation.

T4D11 What organization has published safety guidelines for the maximum limits of RF energy near the human body?
A. The Institute of Electrical and Electronics Engineers (IEEE)
B. The Federal Communications Commission (FCC)
C. The Environmental Protection Agency (EPA)
D. The American National Standards Institute (ANSI)

ANSWER D: This organization is the authority on RF protection. You will see their acronym, ANSI, often on standards and specifications.

T4D12 What is the purpose of the ANSI RF protection guide?
- A. It lists all RF frequency allocations for interference protection.
- B. It gives RF exposure limits for the human body.
- C. It sets transmitter power limits for interference protection.
- D. It sets antenna height limits for aircraft protection.

ANSWER B: Watch out when you are working with antennas. RF protection is essential.

T4D13 According to the ANSI RF protection guide, what frequencies cause us the greatest risk from RF energy?
- A. 3 to 30 MHz
- B. 300 to 3000 MHz
- C. Above 1500 MHz
- D. 30 to 300 MHz

ANSWER D: When you get up to the VHF bands, exposure to RF gets to be a big issue.

T4D14 Why is the limit of exposure to RF the lowest in the frequency range of 30 MHz to 300 MHz, according to the ANSI RF protection guide?
- A. There are more transmitters operating in this range.
- B. There are fewer transmitters operating in this range.
- C. Most transmissions in this range are for a longer time.
- D. The human body absorbs RF energy the most in this range.

ANSWER D: Within our body, we have many small organs and muscles with a length that corresponds to VHF wavelength frequencies. This is why we must be concerned with RF exposure between 30 MHz and 300 MHz—our body parts are actually resonant!

T4D15 According to the ANSI RF protection guide, what is the maximum safe power output to the antenna of a hand-held VHF or UHF radio?
- A. 125 milliwatts
- B. 7 watts
- C. 10 watts
- D. 25 watts

ANSWER B: Most VHF and UHF hand-helds don't put out much over 5 watts, which is 2 watts under the 7-watt limit.

T4D16 After you have opened a VHF power amplifier to make internal tuning adjustments, what should you do before you turn the amplifier on?
- A. Remove all amplifier shielding to ensure maximum cooling.
- B. Make sure that the power interlock switch is bypassed so you can test the amplifier.
- C. Be certain all amplifier shielding is fastened in place.
- D. Be certain no antenna is attached so that you will not cause any interference.

ANSWER C: Transmitter shielding is important to prevent unwanted radiation of RF signals to your body.

Subelement T5 – Electrical Principles

2 exam questions
2 topic groups

T5A Definition of resistance, inductance, and capacitance and unit of measurement, calculation of values in series and parallel

T5A01 What does resistance do in an electric circuit?
A. It stores energy in a magnetic field.
B. It stores energy in an electric field.
C. It provides electrons by a chemical reaction.
D. It opposes the flow of electrons.

ANSWER D: Think of resistance as stepping on a garden hose. This restricts the water flow, similar to a resistor that restricts the flow of electrons in a direct current (dc) circuit.

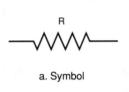

R

a. Symbol

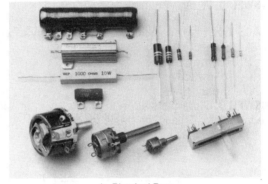

b. Physical Part

Resistor

T5A02 What is the ability to store energy in a magnetic field called?
A. Admittance
B. Capacitance
C. Resistance
D. Inductance

ANSWER D: Place a magnetic compass near an energized inductor from dc and watch what happens! The compass needle lines up with the magnetic field, which is the basis for inductance.

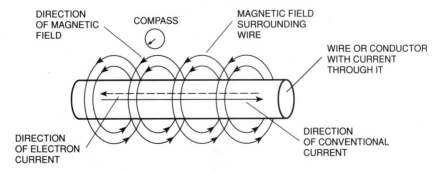

DIRECTION OF MAGNETIC FIELD

COMPASS

MAGNETIC FIELD SURROUNDING WIRE

WIRE OR CONDUCTOR WITH CURRENT THROUGH IT

DIRECTION OF ELECTRON CURRENT

DIRECTION OF CONVENTIONAL CURRENT

Magnetic Field Around a Conductor Carrying Current

3

T5A03 What is the basic unit of inductance?
A. The coulomb
B. The farad
C. The henry
D. The ohm

ANSWER C: It is the henry, but because the henry is a fairly large unit, we measure inductance in one thousandths of a henry (millihenry) and one millionths of a henry (microhenry).

T5A04 What is a henry?
A. The basic unit of admittance
B. The basic unit of capacitance
C. The basic unit of inductance
D. The basic unit of resistance

ANSWER C: When you think of inductance, think of the henry.

T5A05 What is the ability to store energy in an electric field called?
A. Inductance
B. Resistance
C. Tolerance
D. Capacitance

ANSWER D: Capacitors store their energy in an electric field, not a magnetic field like inductors. This is the basis for capacitance.

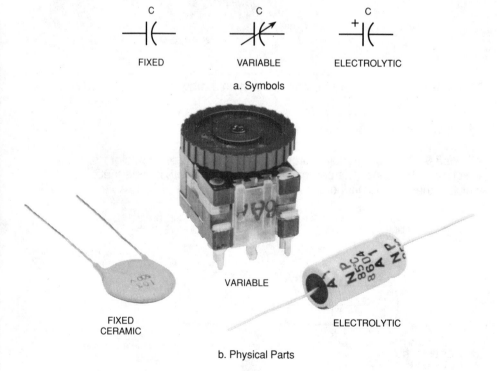

a. Symbols

b. Physical Parts

Capacitors

T5A06 What is the basic unit of capacitance?
 A. The farad
 B. The ohm
 C. The volt
 D. The henry
ANSWER A: It is the farad, but because the farad is a fairly large unit, we measure capacitance in one millionths of a farad (microfarad) or one million millionths of a farad (picofarad).

T5A07 What is a farad?
 A. The basic unit of resistance
 B. The basic unit of capacitance
 C. The basic unit of inductance
 D. The basic unit of admittance
ANSWER B: When you think of capacitance, remember the farad.

T5A08 If two equal-value inductors are connected *in series,* what is their total inductance?
 A. Half the value of one inductor
 B. Twice the value of one inductor
 C. The same as the value of either inductor
 D. The value of one inductor times the value of the other
ANSWER B: Inductors are treated like resistors with respect to total value calculations. Their values add in series.

$$L_1 + L_2 = L_T$$
$$\text{When } L_1 = L_2, \quad L_T = 2L_2$$

T5A09 If two equal-value inductors are connected *in parallel,* what is their total inductance?
 A. Half the value of one inductor
 B. Twice the value of one inductor
 C. The same as the value of either inductor
 D. The value of one inductor times the value of the other
ANSWER A: Inductors in parallel are treated just like resistors in parallel with respect to total value calculations. Total value is less than value of either one.

$$\frac{L_1 \times L_2}{L_1 + L_2} = L_T$$
$$\text{When } L_1 = L_2, \quad L_T = \frac{L_2^2}{2L} = \frac{L_2}{2}$$

T5A10 If two equal-value capacitors are connected *in series,* what is their total capacitance?
 A. Twice the value of one capacitor
 B. The same as the value of either capacitor
 C. Half the value of either capacitor
 D. The value of one capacitor times the value of the other
ANSWER C: Capacitor values combine just opposite of resistors. The total value is less than value of either one.The general equation for finding the equivalent capacitance of any two capacitors in series is:

$$\frac{C_1 \times C_2}{C_1 + C_2} = C_T$$

If $\qquad C_2 = C_1 \quad$ Then $\dfrac{C_1 \times C_1}{C_1 + C_1} = C_T$

Therefore, $\dfrac{C_1{}^2}{2C_1} = C_T \quad$ And $\qquad \dfrac{C_1}{2} = C_T$

Makes Smaller Capacitor

Series Capacitance Equation

T5A11 If two equal-value capacitors are connected *in parallel*, what is their total capacitance?
A. Twice the value of one capacitor
B. Half the value of one capacitor
C. The same as the value of either capacitor
D. The value of one capacitor times the value of the other

ANSWER A: Treat capacitors just the opposite of what you would with resistors. Capacitors in parallel add up to make larger value capacitors.

$$C_1 + C_2 = C_T$$

If $\qquad C_2 = C_1$

Then, $\qquad C_1 + C_1 = C_T$

Therefore, $\qquad 2C_1 = C_T$

Makes Larger Capacitor

Parallel Capacitance Equation

T5B Ohm's Law

T5B01 Ohm's Law describes the mathematical relationship between what three electrical quantities?
A. Resistance, voltage and power
B. Current, resistance and power
C. Current, voltage and power
D. Resistance, current and voltage

ANSWER D: Ohm's Law describes the relationship between voltage (E), current (I), and resistance (R). If you draw this magic circle, it's easy to remember how to do calculations for any question using Ohms Law:

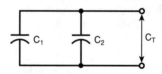

For Voltage: $\quad E = I \times R$

For Current: $\quad I = \dfrac{E}{R}$

For Current: $\quad R = \dfrac{E}{I}$

E = VOLTAGE IN VOLTS
I = CURRENT IN AMPERES
R = RESISTANCE IN OHMS

Ohm's Law Calculation

The magic circle shows E, I and R in position so it provides the correct equation for your problem. To use the magic circle, cover with your finger the letter for which you are solving. Now plug in the other two values that they give you in the examination question. Solve the problem by performing the mathematical operation indicated by the position of the remaining letters.

T5B02 How is the current in a DC circuit calculated when the voltage and resistance are known?
A. I = R × E [current equals resistance multiplied by voltage]
B. I = R / E [current equals resistance divided by voltage]
C. I = E / R [current equals voltage divided by resistance]
D. I = P / E [current equals power divided by voltage]
ANSWER C: Use the magic circle from question T5B01. In this question, solve for I (current) which is equal to E (voltage) divided by R (resistance).

T5B03 How is the resistance in a DC circuit calculated when the voltage and current are known?
A. R = I / E [resistance equals current divided by voltage]
B. R = E / I [resistance equals voltage divided by current]
C. R = I × E [resistance equals current multiplied by voltage]
D. R = P / E [resistance equals power divided by voltage]
ANSWER B: Use the Ohm's Law magic circle in T5B01. Put your finger over the letter "R". What you have left is "E" over "I", or voltage divided by current.

T5B04 How is the voltage in a DC circuit calculated when the current and resistance are known?
A. E = I / R [voltage equals current divided by resistance]
B. E = R / I [voltage equals resistance divided by current]
C. E = I × R [voltage equals current multiplied by resistance]
D. E = P / I [voltage equals power divided by current]
ANSWER C: Now we are looking for voltage. Put your finger over the letter "E" in the Ohm's Law magic circle in T5B01. This leaves the letter "I" and the letter "R" on the same line. This means voltage equals current multiplied by resistance. See how simple this is.

T5B05 If a 12-volt battery supplies 0.25 ampere to a circuit, what is the circuit's resistance?
A. 0.25 ohm
B. 3 ohm
C. 12 ohms
D. 48 ohms
ANSWER D: This question asks for the resistance. Put your finger over R in the magic circle, and substitute the given numbers for the letters. Substitute 12 for E and 0.25 for I. Since R is equal to E over I, R is now equal to 12 divided by 0.25 or R = 12/.25 You may do this longhand or you may use a calculator on the Technician test. Your answer should come out 48 ohms. Here are the calculator keystrokes: CLEAR 12 ÷ .25 = and the answer is 48.

T5B06 If a 12-volt battery supplies 0.15 ampere to a circuit, what is the circuit's resistance?
A. 0.15 ohm
B. 1.8 ohm

C. 12 ohms

D. 80 ohms

ANSWER D: Refer to question TB501. Write down that Ohm's Law circle! Learn it well and be able to apply it easily. This problem is similar to question T5B05, solving for R, but in this case the current is only 0.15 ampere. R is equal to 12 divided by I which equals 0.15, or R = 12/.15. Your answer is 80 ohms. The calculator keystrokes are: CLEAR 12 ÷ .15 = and the answer is 80.

T5B07 If a 4800-ohm resistor is connected to 120 volts, approximately how much current will flow through it?

A. 4 A

B. 25 mA

C. 25 A

D. 40 MA

ANSWER B: We are solving for current. "I" is equal to "E" divided by "R". Calculators are okay for the test. The calculator keystrokes are: CLEAR 120 ÷ 4800 = and the answer is 0.025 amps. Move the decimal point 3 places to the right to convert 0.025 amps to 25 milliamps (mA), so answer choice B is the correct answer.

T5B08 If a 48000-ohm resistor is connected to 120 volts, approximately how much current will flow through it?

A. 400 A

B. 40 A

C. 25 mA

D. 2.5 mA

ANSWER D: The calculator keystrokes are: CLEAR 120 ÷ 48000 = and your answer is 0.0025 amps. Move the decimal point 3 places to the right and the correct answer is 2.5 mA. See how simple this is? Let's try one more.

T5B09 If a 4800-ohm resistor is connected to 12 volts, approximately how much current will flow through it?

A. 2.5 mA

B. 25 mA

C. 40 A

D. 400 A

ANSWER A: The calculator keystrokes are: CLEAR 12 ÷ 4800 = and your answer is 0.0025 amps. Move the decimal point 3 places to the right for the correct answer of 2.5 mA. Okay, you talked me into it—one more.

T5B10 If a 48000-ohm resistor is connected to 12 volts, approximately how much current will flow through it?

A. 250 uA

B. 250 mA

C. 4000 mA

D. 4000 A

ANSWER A: You know the steps: CLEAR 12 ÷ 48000 =. The answer is 0.00025 amps. Moving the decimal 3 places to the right gives 0.25 ma but that is not an answer choice. So move the decimal three more places to the right (total of 6 places) and you have 250 microamps (µA) which is an answer choice. Remember a microampere equals one millionth (1 × 10⁻⁶) of an ampere and the symbol is µA.

T5B11 If you know the voltage and current supplied to a circuit, what formula would you use to calculate the circuit's resistance?
- A. Ohm's law
- B. Tesla's law
- C. Ampere's law
- D. Kirchhoff's law

ANSWER A: It is Ohm's Law that allows you to calculate circuit resistance when you know the voltage and current supplied to that circuit. Every time you have an Ohm's Law problem, always put down the magic circle. Then place your finger over what they are looking for, and the rest will be easy with a calculator. All examiners permit the use of calculators during the examination. Just be careful when entering the values so you input the correct number of zeros.

Subelement T6 – Circuit Components	2 exam questions
	2 topic groups

T6A Resistors, construction types, variable and fixed, color code, power ratings, schematic symbols

T6A01 Which of the following are the most common resistor types?
- A. Plastic and porcelain
- B. Film and wire-wound
- C. Electrolytic and metal-film
- D. Iron core and brass core

ANSWER B: File these resistor types in your memory. See question T5A01.

T6A02 What does a variable resistor or potentiometer do?
- A. Its resistance changes when AC is applied to it.
- B. It transforms a variable voltage into a constant voltage.
- C. Its resistance changes when its slide or contact is moved.
- D. Its resistance changes when it is heated.

ANSWER C: The volume control on your worldwide radio is a variable resistor. (Be careful: Answer D is also technically correct, but it is not the answer they want.)

Variable Resistor

T6A03 How do you find a resistor's tolerance rating?
- A. By using a voltmeter
- B. By reading the resistor's color code

C. By using Thevenin's theorem for resistors

D. By reading its Baudot code

ANSWER B: It's easy to read a resistor's color code. The first three bands indicate the resistor's resistance. The first color indicates the first number, the second color the second number, and the third color indicates the number of zeros to add after the first two numbers. For all three bands, black is zero, brown is one, red is two, orange is three, yellow is four, green is five, blue is six, violet is seven, gray is eight, and white is nine. Thus, red/black/red indicates 2-0-00 for a value of 2000 ohms. The fourth band indicates the permitted tolerance of the indicated nominal value. If the fourth band is gold, the tolerance is good at ±5 percent. Silver is an okay ±10 percent tolerance; and if there is no fourth band, the tolerance is assumed to be a ±20 percent average resistor. Thus, yellow/violet/orange/gold indicates a nominal value of 47000 ohms with a possible range from 44650 to 49350 ohms.

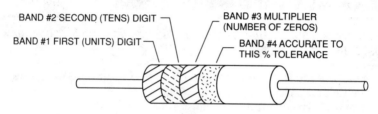

BAND #2 SECOND (TENS) DIGIT

BAND #1 FIRST (UNITS) DIGIT

BAND #3 MULTIPLIER (NUMBER OF ZEROS)

BAND #4 ACCURATE TO THIS % TOLERANCE

EXAMPLES

Band	1	2	3	4	Value
Resistor #1	Red (2)	Black (0)	Red (00)	Gold 5%	2,000 ±5%
Resistor #2	Brown (1)	Black (0)	Green (00000)	—	1,000,000 ±20%
Resistor #3	Blue (6)	White (9)	Orange (000)	Silver 10%	69,000 ±10%

Resistor Color Code (First Three Bands)

Black	0	Blue	6
Brown	1	Violet	7
Red	2	Gray	8
Orange	3	White	9
Yellow	4	Silver	0.01
Green	5	Gold	0.1

Tolerance (Fourth Band)

Gold	±5%
Silver	±10%
None	±20%

Resistor Color Code

Source: *Technology Dictionary* © 1987, Master Publishing, Inc., Richardson, Texas

---

T6A04 What do the first three color bands on a resistor indicate?

A. The value of the resistor in ohms

B. The resistance tolerance in percent

C. The power rating in watts

D. The resistance material

ANSWER A: You read the ohmic value of a resistor by the color of the bands. See question T6A03.

T6A05 What does the fourth color band on a resistor indicate?

A. The value of the resistor in ohms

B. The resistance tolerance in percent

C. The power rating in watts

D. The resistance material

ANSWER B: The fourth color band indicates the tolerance. No fourth color band means ±20 percent. See question T6A03.

T6A06 Why do resistors sometimes get hot when in use?

A. Some electrical energy passing through them is lost as heat.

B. Their reactance makes them heat up.

C. Hotter circuit components nearby heat them up.

D. They absorb magnetic energy which makes them hot.

ANSWER A: Resistors get hot as they burn up energy.

T6A07 Why would a large size resistor be used instead of a smaller one of the same resistance?

A. For better response time

B. For a higher current gain

C. For greater power dissipation

D. For less impedance in the circuit

ANSWER C: Besides the actual resistance and tolerance of a resistor, its power dissipation plays an important part in resistor selection. Physically larger resistors can dissipate more power than smaller resistors with the same resistance value.

T6A08 What are the possible values of a 100-ohm resistor with a 10% tolerance?

A. 90 to 100 ohms

B. 10 to 100 ohms

C. 90 to 110 ohms

D. 80 to 120 ohms

ANSWER C: If a 100-ohm resistor has a ±10 percent tolerance, it could be 10 ohms less or 10 ohms more than 100 ohms; that is, its measured value could range from 90 ohms to 110 ohms.

T6A09 How do you find a resistor's value?

A. By using a voltmeter

B. By using the resistor's color code

C. By using Theorem's theorem for resistors

D. By using the Baudot code

ANSWER B: For the Technician Class examination, you do not need to memorize the resistor color code. However, to be a good Amateur Radio experimenter, it's handy to have it memorized. See question T6A03.

T6A10 Which tolerance rating would a high-quality resistor have?

A. 0.1%

B. 5%

C. 10%

D. 20%

ANSWER A: In a precision circuit that you are building, you may need an very high-quality resistor, and a 0.1 percent tolerance resistor would be just fine. High-quality, high-precision resistors cost more and often are physically larger for the same power rating as a low-precision resistor.

T6A11 Which tolerance rating would a low-quality resistor have?

A. 0.1%
B. 5%
C. 10%
D. 20%

ANSWER D: For some circuits, the resistor value may not need to be very close to its nominal value, so ±20 percent tolerance would be okay. Low quality resistors cost less.

T6B Schematic symbols - inductors and capacitors, construction of variable and fixed, factors affecting inductance and capacitance, capacitor construction

T6B01 What is an inductor core?

A. The place where a coil is tapped for resonance
B. A tight coil of wire used in a transformer
C. Insulating material placed between the wires of a transformer
D. The place inside an inductor where its magnetic field is concentrated

ANSWER D: The iron cores used as inductor cores of rf coils are usually very fragile and brittle. If you must adjust them, use the proper tool and adjust them carefully. A metal screwdriver can damage the core and also change its magnetic properties so you cannot make the adjustment properly. A plastic or fiber non-magnetic tool that fits the adjustment slot properly is recommended.

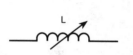

a. Symbol for Variable Inductor

b. Physical Part

Inductor

T6B02 What does an inductor do?

A. It stores a charge electrostatically and opposes a change in voltage.
B. It stores a charge electrochemically and opposes a change in current.
C. It stores a charge electromagnetically and opposes a change in current.
D. It stores a charge electromechanically and opposes a change in voltage.

ANSWER C: Remember that coils develop a magnetic field which is indicated by a magnetic compass held near the energized coil. Energy is stored in the field.

T6B03 What determines the inductance of a coil?

A. The core material, the core diameter, the length of the coil and whether the coil is mounted horizontally or vertically
B. The core diameter, the number of turns of wire used to wind the coil and the type of metal used for the wire

---

C. The core material, the number of turns used to wind the core and the frequency of the current through the coil

D. The core material, the core diameter, the length of the coil and the number of turns of wire used to wind the coil

ANSWER D: All these factors determine the inductance of a coil. An iron core within the coil increases inductance; a brass core reduces inductance.

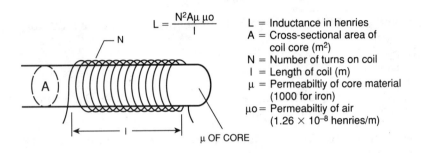

$$L = \frac{N^2 A \mu \, \mu o}{l}$$

L = Inductance in henries
A = Cross-sectional area of coil core (m²)
N = Number of turns on coil
l = Length of coil (m)
μ = Permeabiltiy of core material (1000 for iron)
μo = Permeabiltiy of air (1.26 × 10⁻⁸ henries/m)

μ OF CORE

Inductance Equation

T6B04 As an iron core is inserted in a coil, what happens to the coil's inductance?

A. It increases.
B. It decreases.
C. It stays the same.
D. It disappears.

ANSWER A: Never adjust the iron core in a coil unless you know exactly what you are doing. They are adjusted properly at the factory and wax-sealed so they seldom vibrate loose. Usually an adjustment is not necessary unless a repair has changed the circuit's characteristics.

T6B05 What can happen if you tune a ferrite-core coil with a metal tool?

A. The metal tool can change the coil's inductance and cause you to tune the coil incorrectly.
B. The metal tool can become magnetized so much that you might not be able to remove it from the coil.
C. The metal tool can pick up enough magnetic energy to become very hot.
D. The metal tool can pick up enough magnetic energy to become a shock hazard.

ANSWER A: Always use a plastic adjustment tool. A metal tool can change the coil's inductance and can physically damage the core.

T6B06 In Figure T6-1 which symbol represents an adjustable inductor?

A. Symbol 1
B. Symbol 2
C. Symbol 3
D. Symbol 4

ANSWER B: The arrow indicates the inductor is adjustable.

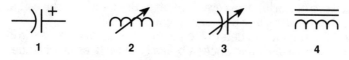

Figure T6-1

T6B07 In Figure T6-1 which symbol represents an iron-core inductor?
A. Symbol 1
B. Symbol 2
C. Symbol 3
D. Symbol 4
ANSWER D: The two lines above the coil indicate an iron core. Don't confuse the coil symbol with a resistor symbol.

T6B08 In Figure T6-1 which symbol represents an inductor wound over a toroidal core?
A. Symbol 1
B. Symbol 2
C. Symbol 3
D. Symbol 4
ANSWER D: There is no difference in the way a toroidal core is shown on a schematic diagram.

T6B09 In Figure T6-1 which symbol represents an electrolytic capacitor?
A. Symbol 1
B. Symbol 2
C. Symbol 3
D. Symbol 4
ANSWER A: Symbols 2 and 4 are coils, and symbol 3 is an adjustable capacitor. Symbol 1 is the electrolytic capacitor. Notice the "plus" sign that indicates polarity. It must be connected in the crcuit with correct polarity.

T6B10 In Figure T6-1 which symbol represents a variable capacitor?
A. Symbol 1
B. Symbol 2
C. Symbol 3
D. Symbol 4
ANSWER C: Symbol 2 is an adjustable coil. You want the adjustable capacitor, so symbol 3 is the correct answer.

T6B11 What describes a capacitor?
A. Two or more layers of silicon material with an insulating material between them
B. The wire used in the winding and the core material
C. Two or more conductive plates with an insulating material between them
D. Two or more insulating plates with a conductive material between them
ANSWER C: The plates never touch. See questions T6B13 and T5A05.

Variable Capacitor

T6B12 What does a capacitor do?
A. It stores a charge electrochemically and opposes a change in current.
B. It stores a charge electrostatically and opposes a change in voltage.
C. It stores a charge electromagnetically and opposes a change in current.
D. It stores a charge electromechanically and opposes a change in voltage.

ANSWER B: Capacitors store their energy in an electric field, not a magnetic field, as in a coil.

T6B13 What determines the capacitance of a capacitor?
A. The material between the plates, the area of one side of one plate, the number of plates and the spacing between the plates
B. The material between the plates, the number of plates and the size of the wires connected to the plates
C. The number of plates, the spacing between the plates and whether the dielectric material is N type or P type
D. The material between the plates, the area of one plate, the number of plates and the material used for the protective coating

ANSWER A: Memorize this answer for the examination. Read over the incorrect answers and see why they are wrong. Larger area of plates, larger number of plates, and closer spacing—all increase capacitance.

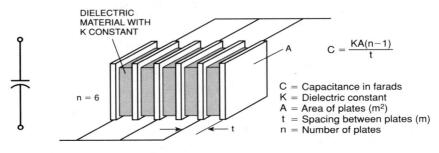

DIELECTRIC MATERIAL WITH K CONSTANT

A

$$C = \frac{KA(n-1)}{t}$$

n = 6

C = Capacitance in farads
K = Dielectric constant
A = Area of plates (m²)
t = Spacing between plates (m)
n = Number of plates

t

a. Symbol

b. Parallel Plate Capacitor (6 Plates)

Capacitance Equation
Source: *Using You Meter*, A.J. Evans, © 1985, Master Publishing, Inc.

T6B14 As the plate area of a capacitor is increased, what happens to its capacitance?
 A. It decreases.
 B. It increases.
 C. It stays the same.
 D. It disappears.
ANSWER B: The larger the plate area, the greater the capacitance.

Subelement T7 – Practical Circuits

1 exam question
1 topic group

T7A Practical circuits

T7A01 Why do modern HF transmitters have a built-in low-pass filter in their RF output circuits?
 A. To reduce RF energy below a cutoff point
 B. To reduce low-frequency interference to other amateurs
 C. To reduce harmonic radiation
 D. To reduce fundamental radiation
ANSWER C: Your worldwide set already has a built-in low-pass filter connected to its output. This usually is fine for reducing harmonics.

T7A02 What circuit blocks RF energy above and below a certain limit?
 A. A band-pass filter
 B. A high-pass filter
 C. An input filter
 D. A low-pass filter
ANSWER A: Read this question carefully! Since it blocks energy above and below a certain frequency, it must be a band-pass filter.

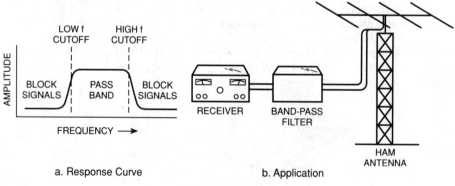

a. Response Curve   b. Application

Band-Pass Filter

T7A03 What type of filter is used in the IF section of receivers to block energy outside a certain frequency range?
 A. A band-pass filter
 B. A high-pass filter
 C. An input filter
 D. A low-pass filter
ANSWER A: Band-pass filters are used in many types of radio receivers.

T7A04 What circuit is found in all types of receivers?
A. An audio filter
B. A beat-frequency oscillator
C. A detector
D. An RF amplifier

ANSWER C: All radio receivers need a circuit to demodulate or detect the information impressed on the rf carrier. A common detector detects the audio signal from the rf carrier.

T7A05 What circuit has a variable-frequency oscillator connected to a driver and a power amplifier?
A. A packet-radio transmitter
B. A crystal-controlled transmitter
C. A single-sideband transmitter
D. A VFO-controlled transmitter

ANSWER D: "VFO" stands for "variable-frequency oscillator", so answer D is correct.

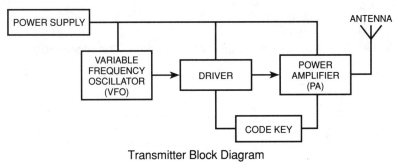

Transmitter Block Diagram

T7A06 What circuit combines signals from an IF amplifier stage and a beat-frequency oscillator (BFO), to produce an audio signal?
A. An AGC circuit
B. A detector circuit
C. A power supply circuit
D. A VFO circuit

ANSWER B: It is the task of the detector circuit to combine signals from an intermediate frequency (IF) amplifier and a beat frequency oscillator to detect (recover) the audio signal.

T7A07 What circuit uses a limiter and a frequency discriminator to pro-duce an audio signal?
A. A double-conversion receiver
B. A variable-frequency oscillator
C. A superheterodyne receiver
D. A FM receiver

ANSWER D: When you see the words "frequency discriminator", you know they talking about a stage of a frequency modulation (FM) receiver.

T7A08 What circuit is pictured in Figure T7-1 if block 1 is a variable-frequency oscillator?
A. A packet-radio transmitter
B. A crystal-controlled transmitter

C. A single-sideband transmitter

D. A VFO-controlled transmitter

ANSWER D: A variable frequency oscillator is abbreviated "VFO". Answer D is correct.

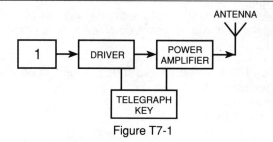

Figure T7-1

T7A09 What is the unlabeled block in Figure T7-2?

A. An AGC circuit

B. A detector

C. A power supply

D. A VFO circuit

ANSWER B: That stage which combines the IF and BFO is a detector.

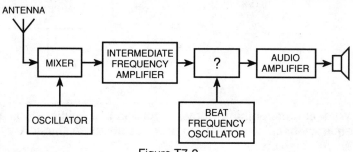

Figure T7-2

T7A10 What circuit is pictured in Figure T7-3?

A. A double-conversion receiver

B. A variable-frequency oscillator

C. A superheterodyne receiver

D. An FM receiver

ANSWER D: Did you spot the frequency discriminator in this block diagram? This means we are looking at a frequency modulation receiver.

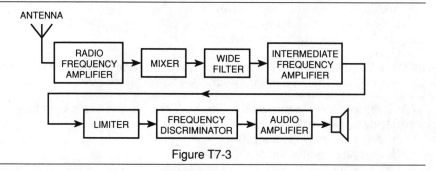

Figure T7-3

T7A11 What is the unlabeled block in Figure T7-4?
A. A band-pass filter
B. A crystal oscillator
C. A reactance modulator
D. A rectifier modulator

ANSWER C: Here we see a transmitter. It is a frequency modulated transmitter which always requires a reactance modulator. Watch out for answer D—it looks close, but answer C—a reactance modulator—is the correct answer.

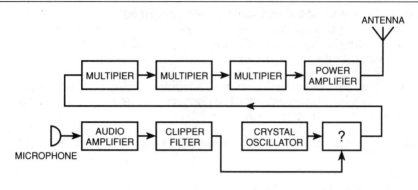

Figure T7-4

Subelement T8 – Signals and Emissions

2 exam questions
2 topic groups

T8A Definition of modulation and emission types

T8A01 What is the name for unmodulated carrier wave emissions?
A. Phone
B. Test
C. CW
D. RTTY

ANSWER B: In the "test" mode, an unmodulated carrier wave has no sidebands.

T8A02 What is the name for Morse code emissions produced by switching a transmitter's output on and off?
A. Phone
B. Test
C. CW
D. RTTY

ANSWER C: Morse code is a form of continuous wave, or CW.

T8A03 What is RTTY?
A. Amplitude-keyed telegraphy
B. Frequency-shift-keyed telegraphy
C. Frequency-modulated telephony
D. Phase-modulated telephony

ANSWER B: Frequency-Shift-Keying (FSK) is the same as RTTY on worldwide frequencies.

T8A04 What is the name for packet-radio emissions?

A. CW
B. Data
C. Phone
D. RTTY

ANSWER B: Another name for sending high-speed digital packet radio emissions is called "data". CW is code, phone is voice, and RTTY is a rather slow radioteleprinter.

T8A05 How is tone-modulated Morse code produced?

A. By feeding a microphone's audio signal into an FM transmitter
B. By feeding an on/off keyed audio tone into a CW transmitter
C. By on/off keying of a carrier
D. By feeding an on/off keyed audio tone into a transmitter

ANSWER D: It would be impractical to send telegraphy using the touch-tone pad of your handie-talkie. It can be done, but don't do it—it gets your transmitter hot and consumes a lot of battery power.

T8A06 What is the name of the voice emission most used on VHF/UHF repeaters?

A. Single-sideband phone
B. Pulse-modulated phone
C. Slow-scan phone
D. Frequency-modulated phone

ANSWER D: We use frequency modulation on most VHF and UHF repeaters.

T8A07 What is the name of the voice emission most used on amateur HF bands?

A. Single-sideband phone
B. Pulse-modulated phone
C. Slow-scan phone
D. Frequency modulated phone

ANSWER A: On the worldwide bands, frequency modulation is too wide. For phone, we always use single sideband. On 40 through 160 meters, we use lower sideband; on 10 through 20 meters, we use upper sideband. See figure at question T8A08.

T8A08 What is meant by the upper-sideband (USB)?

A. The part of a single-sideband signal which is above the carrier frequency
B. The part of a single-sideband signal which is below the carrier frequency
C. Any frequency above 10 MHz
D. The carrier frequency of a single-sideband signal

ANSWER A: USB stands for upper sideband, found on worldwide bands 10 through 20 meters. Lower sideband is found on 40 meters through 160 meters.

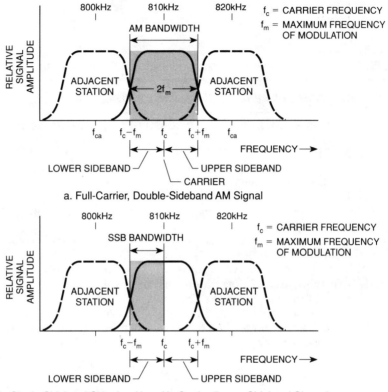

a. Full-Carrier, Double-Sideband AM Signal

b. Single-Sideband Signal — Uses No Carrier (Lower Sideband Shown)

Carrier, Sidebands and Bandwidth

Source: *Mobile 2-Way Radio Communications,* G.West, © 1993, Master Publishing, Inc.

T8A09 What emissions are produced by a transmitter using a reactance modulator?
A. CW
B. Test
C. Single-sideband, suppressed-carrier phone
D. Phase-modulated phone

ANSWER D: Some VHF transceivers use phase modulation, as opposed to frequency modulation. They sound a little bit different on the air.

T8A10 What other emission does phase modulation most resemble?
A. Amplitude modulation
B. Pulse modulation
C. Frequency modulation
D. Single-sideband modulation

ANSWER C: Frequency modulation and phase modulation sound almost the same unless you listen very carefully.

T8A11 What is the name for emissions produced by an on/off keyed audio tone?

    A. RTTY
    B. MCW
    C. CW
    D. Phone

ANSWER B: This is Morse code practice using an FM transmitter.

T8B RF carrier, modulation, bandwidth and deviation

T8B01 What is another name for a constant-amplitude radio-frequency signal?

    A. An RF carrier
    B. An AF carrier
    C. A sideband carrier
    D. A subcarrier

ANSWER A: The carrier is a pure signal, without tone (unmodulated). See figure at question T8B02.

T8B02 What is modulation?

    A. Varying a radio wave in some way to send information
    B. Receiving audio information from a signal
    C. Increasing the power of a transmitter
    D. Suppressing the carrier in a single-sideband transmitter

ANSWER A: When we modulate a transmitter, we usually impress voice or audio tones on the carrier wave.

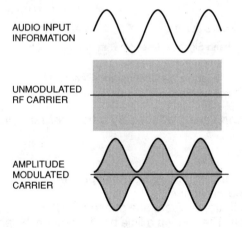

AUDIO INPUT INFORMATION

UNMODULATED RF CARRIER

AMPLITUDE MODULATED CARRIER

Amplitude Modulation

T8B03 What kind of emission would your FM transmitter produce if its microphone failed to work?

    A. An unmodulated carrier
    B. A phase-modulated carrier
    C. An amplitude-modulated carrier
    D. A frequency-modulated carrier

ANSWER A: If the microphone doesn't work, then there is no audio to modulate the carrier, thus the emission would be an unmodulated carrier.

T8B04 How would you modulate a 2-meter FM transceiver to produce packet-radio emissions?
A. Connect a terminal-node-controller to interrupt the transceiver's carrier wave.
B. Connect a terminal-node-controller to the transceiver's microphone input.
C. Connect a keyboard to the transceiver's microphone input.
D. Connect a DTMF key pad to the transceiver's microphone input.

ANSWER B: For packet radio use, you will need a terminal-node controller that is hooked into the transceiver's microphone input. On some sets, an input jack for the controller is provided on the back.

T8B05 Why is FM voice best for local VHF/UHF radio communications?
A. The carrier is not detectable.
B. It is more resistant to distortion caused by reflected signals.
C. It has high-fidelity audio which can be understood even when the signal is somewhat weak.
D. Its RF carrier stays on frequency better than the AM modes.

ANSWER C: FM is good even when signals are weak. Since most FM work is line-of-sight, FM is a delightful mode to use on frequencies above 50 MHz.

T8B06 Why do many radio receivers have several IF filters of different bandwidths that can be selected by the operator?
A. Because some frequency bands are wider than others
B. Because different bandwidths help increase the receiver sensitivity
C. Because different bandwidths improve S-meter readings
D. Because some emission types need a wider bandwidth than others to be received properly

ANSWER D: A better worldwide transceiver has several different IF filters that may be selected. Look for this feature when selecting your worldwide base station.

T8B07 Which list of emission types is in order from the narrowest bandwidth to the widest bandwidth?
A. RTTY, CW, SSB voice, FM voice
B. CW, FM voice, RTTY, SSB voice
C. CW, RTTY, SSB voice, FM voice
D. CW, SSB voice, RTTY, FM voice

ANSWER C: This gives you a clue why FM voice frequencies are always found in the VHF and UHF spectrum—and the only place FM is found on worldwide frequencies is between 29.5 and 29.7 MHz. Because of FM's wide bandwidth, using FM below VHF frequencies would exceed bandwidth limitations.

T8B08 What is the usual bandwidth of a single-sideband amateur signal?
A. 1 kHz
B. 2 kHz
C. Between 3 and 6 kHz
D. Between 2 and 3 kHz

ANSWER D: A single-sideband amateur radio signal is typically 2 to 3 kHz wide. This is half of what an original double-sideband signal was in the old days.

T8B09 What is the usual bandwidth of a frequency-modulated amateur signal?
A. Less than 5 kHz
B. Between 5 and 10 kHz
C. Between 10 and 20 kHz
D. Greater than 20 kHz

ANSWER C: When we transmit using frequency modulated phone (FM), our signal swings back and forth (deviates) up to ±5000 Hz. This is a total excursion of 10 kHz. Watch out for answer A—it's wrong because it does not have a $\pm$ in front of it. Answer B is close, but answer C has been designated the correct answer.

T8B10 What is the result of overdeviation in an FM transmitter?
A. Increased transmitter power
B. Out-of-channel emissions
C. Increased transmitter range
D. Poor carrier suppression

ANSWER B: If you overdrive the audio stage of an FM transceiver, you will create over-deviation that creates interference to adjacent channel users.

T8B11 What causes splatter interference?
A. Keying a transmitter too fast
B. Signals from a transmitter's output circuit are being sent back to its input circuit.
C. Overmodulation of a transmitter
D. The transmitting antenna is the wrong length.

ANSWER C: Back off the microphone if someone indicates your signal is "splattering." On some FM sets, you can turn down the modulation gain in the mike circuit. On other sets, you simply turn down the amount of deviation on the deviation pot, which is inside the transmitter.

Subelement T9 – Antennas and Feed Lines

3 exam questions
3 topic groups

T9A Parasitic beam and non-directional antennas

T9A01 What is a directional antenna?
A. An antenna which sends and receives radio energy equally well in all directions
B. An antenna that cannot send and receive radio energy by skywave or skip propagation
C. An antenna which sends and receives radio energy mainly in one direction
D. An antenna which sends and receives radio energy equally well in two opposite directions

ANSWER C: Good examples of directional antennas are quads, rhombics, and the Yagi. See question T9A02 for the Yagi.

T9A02 How is a Yagi antenna constructed?

A. Two or more straight, parallel elements are fixed in line with each other.

B. Two or more square or circular loops are fixed in line with each other.

C. Two or more square or circular loops are stacked inside each other.

D. A straight element is fixed in the center of three or more elements which angle toward the ground.

ANSWER A: The Yagi antenna is really a series of dipoles, all lined up to increase the signal gain in one direction. A dipole is usually a half-wavelength long. However, on the Yagi beam, some are longer and some are shorter than the driven dipole. The ones called reflectors are slightly longer than the driven element, and the ones called directors are slightly shorter. The combination makes the Yagi antenna very directional.

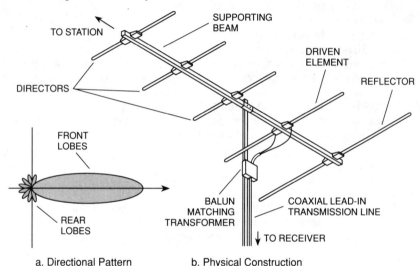

a. Directional Pattern   b. Physical Construction

A Beam Antenna — The Yagi Antenna

Source: *Antennas – Selection and Installation,* © 1986, Master Publishing, Inc., Richardson, Texas

---

T9A03 What type of beam antenna uses two or more straight elements arranged in line with each other?

A. A delta loop antenna

B. A quad antenna

C. A Yagi antenna

D. A Zepp antenna

ANSWER C: A Yagi antenna is a beam antenna.

T9A04 How many directly driven elements do most beam antennas have?

A. None

B. One

C. Two

D. Three

ANSWER B: For the examination, consider only one element as a driven element. However, in the real world of Amateur Radio, there are several beam manufacturers that drive more than one element.

T9A05 What is a parasitic beam antenna?
A. An antenna where some elements obtain their radio energy by induction or radiation from a driven element
B. An antenna where wave traps are used to magnetically couple the elements
C. An antenna where all elements are driven by direct connection to the feed line
D. An antenna where the driven element obtains its radio energy by induction or radiation from director elements

ANSWER A: A beam antenna is an inexpensive investment for excellent power output and extraordinarily good receiving capabilities. It's much better to buy a beam antenna than a power amplifier.

T9A06 What are the parasitic elements of a Yagi antenna?
A. The driven element and any reflectors
B. The director and the driven element
C. Only the reflectors (if any)
D. Any directors or any reflectors

ANSWER D: A parasitic element on a Yagi does not have any coaxial cable attached to it. This is why directors and reflectors are parasitic elements.

T9A07 What is a cubical quad antenna?
A. Four straight, parallel elements in line with each other, each approximately 1/2-electrical wavelength long
B. Two or more parallel four-sided wire loops, each approximately one-electrical wavelength long
C. A vertical conductor 1/4-electrical wavelength high, fed at the bottom
D. A center-fed wire 1/2-electrical wavelength long

ANSWER B: You can add reflectors and directors to your quad antenna system to give it some real punch in just one direction. Just like the Yagi, the reflector element is slightly larger, and the director loop is slightly smaller.

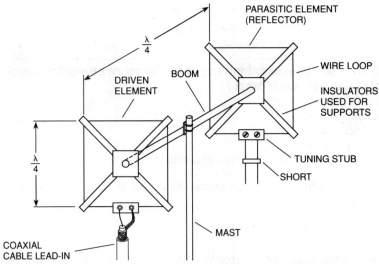

A Two-Element Cubical Quad Antenna — Horizontally Polarized

Source: *Antennas – Selection and Installation*, © 1986, Master Publishing, Inc., Richardson, Texas

T9A08 What is a delta loop antenna?

A. A type of cubical quad antenna, except with triangular elements rather than square

B. A large copper ring or wire loop, used in direction finding

C. An antenna system made of three vertical antennas, arranged in a triangular shape

D. An antenna made from several triangular coils of wire on an insulating form

ANSWER A: The delta loop is a great one to string in trees. It's one complete wavelength, with each side one-third wavelength long.

T9A09 What type of non-directional antenna is easy to make at home and works well outdoors?

A. A Yagi

B. A delta loop

C. A cubical quad

D. A ground plane

ANSWER D: A ground plane antenna is an easy antenna to build. It is an omni-directional antenna that radiates in all directions. For 2 meters, the elements are only 18 inches long. You can make them out of a coat hanger if you are careful to bend the ends back to eliminate sharp points.

T9A10 What type of antenna is made when a magnetic-base whip antenna is placed on the roof of a car?

A. A Yagi

B. A delta loop

C. A cubical quad

D. A ground plane

ANSWER D: A simple 2-meter or 440-MHz mobile antenna has a magnetic-base mount. The radiator is the vertical element and your car is the ground plane. It is an omnidirectional vertical antenna that works well.

T9A11 If a magnetic-base whip antenna is placed on the roof of a car, in what direction does it send out radio energy?

A. It goes out equally well in all horizontal directions.

B. Most of it goes in one direction.

C. Most of it goes equally in two opposite directions.

D. Most of it is aimed high into the air.

ANSWER A: The vertical antenna that may be magnetically mounted on your car radiates vertically polarized radio waves that go out in all horizontal directions. In other words, the signals don't go high into the air, but rather they go out in all horizontal directions to the sides.

T9B Polarization, impedance matching and SWR, feed lines, balanced vs unbalanced (including baluns)

T9B01 What does horizontal wave polarization mean?

A. The magnetic lines of force of a radio wave are parallel to the earth's surface.

B. The electric lines of force of a radio wave are parallel to the earth's surface.

C. The electric lines of force of a radio wave are perpendicular to the earth's surface.

D. The electric and magnetic lines of force of a radio wave are perpendicular to the earth's surface.

ANSWER B: The key words are "electric lines." For local VHF work, make sure that your antenna is polarized in the same plane as your friend's. Antenna polarization is determined by the direction of the electric field of the antenna.

T9B02 What does vertical wave polarization mean?

A. The electric lines of force of a radio wave are parallel to the earth's surface.

B. The magnetic lines of force of a radio wave are perpendicular to the earth's surface.

C. The electric lines of force of a radio wave are perpendicular to the earth's surface.

D. The electric and magnetic lines of force of a radio wave are parallel to the earth's surface.

ANSWER C: For worldwide communications, polarization is not critical because skywave refraction will turn them every which way. However, for VHF propagation, almost everyone uses vertical polarization for best base-to-mobile range.

T9B03 What electromagnetic-wave polarization does a Yagi antenna have when its elements are parallel to the earth's surface?

A. Circular

B. Helical

C. Horizontal

D. Vertical

ANSWER C: If the Yagi elements are parallel with the earth's surface, the electromagnetic wave polarization will be horizontal because the electric field lines are parallel or horizontal to the earth's surface.

T9B04 What electromagnetic-wave polarization does a half-wavelength antenna have when it is perpendicular to the earth's surface?

A. Circular

B. Horizontal

C. Parabolical

D. Vertical

ANSWER D: Waves radiated from an antenna that is perpendicular to the surface of the earth are vertically polarized.

T9B05 What electromagnetic-wave polarization does most man-made electrical noise have in the HF and VHF spectrum?

A. Horizontal

B. Left-hand circular

C. Right-hand circular

D. Vertical

ANSWER D: Think of this answer as just backwards to what you might think when looking at your local horizontal power lines. Those horizontal power lines emit vertical electromagnetic waves. Most electrical interference on HF is loudest with vertically polarized antennas.

T9B06 What does standing-wave ratio mean?
- A. The ratio of maximum to minimum inductances on a feed line
- B. The ratio of maximum to minimum resistances on a feed line
- C. The ratio of maximum to minimum impedances on a feed line
- D. The ratio of maximum to minimum voltages on a feed line

ANSWER D: Standing wave ratio (SWR) is the ratio of maximum voltage to minimum voltage along a transmission line. It is also a ratio of the maximum current to minimum current along a transmission line. It also is the ratio of the power fed forward along a transmission line to the power reflected back along a transmission line.

T9B07 What does forward power mean?
- A. The power traveling from the transmitter to the antenna
- B. The power radiated from the top of an antenna system
- C. The power produced during the positive half of an RF cycle
- D. The power used to drive a linear amplifier

ANSWER A: You should strive for maximum forward power and minimum reflected power. This will give you a perfect match.

T9B08 What does reflected power mean?
- A. The power radiated down to the ground from an antenna
- B. The power returned to a transmitter from an antenna
- C. The power produced during the negative half of an RF cycle
- D. The power returned to an antenna by buildings and trees

ANSWER B: If you have high amounts of reflected power, better check out what's happening at the antenna. Your antenna is either cut to the wrong frequency or has an impedance mismatch. Hopefully, your reflected power will always be less than a watt.

T9B09 What happens to radio energy when it is sent through a poor quality coaxial cable?
- A. It causes spurious emissions.
- B. It is returned to the transmitter's chassis ground.
- C. It is converted to heat in the cable.
- D. It causes interference to other stations near the transmitting frequency.

ANSWER C: Experienced hams can tell how good an antenna match is by feeling the coaxial cable feed line after a few minutes of transmitting. A warm feed line indicates high SWR. If your transmitter is getting red hot, that's also an indication of high SWR. This is a condition that needs to be corrected before melt-down.

T9B10 What is an unbalanced line?
- A. Feed line with neither conductor connected to ground
- B. Feed line with both conductors connected to ground
- C. Feed line with one conductor connected to ground
- D. Feed line with both conductors connected to each other

ANSWER C: Coaxial cable is an unbalanced feed line. Your radio waves travel along the center conductor. The grounded outside braid keeps your radio waves in, and interference out.

T9B11 What device can be installed to feed a balanced antenna with an unbalanced feed line?
- A. A balun
- B. A loading coil
- C. A triaxial transformer
- D. A wavetrap

ANSWER A: Since coax is unbalanced, and we want to feed a balanced beam or dipole with coax cable, a transformer must be used to take the unbalanced signal and put it on both sides of the balanced antenna system. This device may be constructed, or purchased. The word "balun" is a short form of "<u>bal</u>anced to <u>un</u>balanced".

T9C Line losses by line type, length and frequency, RF safety

T9C01 What common connector usually joins RG-213 coaxial cable to an HF transceiver?
- A. An F-type cable connector
- B. A PL-259 connector
- C. A banana plug connector
- D. A binding post connector

ANSWER B: The coaxial cable connector for ham radio use on high frequency and VHF bands is called a PL-259. See figure in question T9C02.

T9C02 What common connector usually joins a hand-held transceiver to its antenna?
- A. A BNC connector
- B. A PL-259 connector
- C. An F-type cable connector
- D. A binding post connector

ANSWER A: For hand-held transceivers, we use a BNC connector for the little rubber whip antenna, or for an adapter to change BNC to SO-239. The BNC connector atop your hand-held is fragile, so don't over-stress it.

Common Antenna Connectors

T9C03 Which of these common connectors has the lowest loss at UHF?
- A. An F-type cable connector
- B. A type-N connector
- C. A BNC connector
- D. A PL-259 connector

ANSWER B: On frequencies above 300 MHz, a PL-259 is not a good choice for an antenna connector. We would rather use a type-N connector that has low loss characteristics. See figure in question T9C02.

T9C04 If you install a 6-meter Yagi antenna on a tower 150 feet from your transmitter, which of the following feed lines is best?
  A. RG-213
  B. RG-58
  C. RG-59
  D. RG-174

ANSWER A: RG-213 is the best choice out of the given answers. However, your author recommends 7/8″ hardline, which is the ultimate in low-loss, semi-flexible cable.

Coax Cable Type, Size and Loss per 100 Feet

Coax Type	Size	Loss at HF 10 MHz	Loss at UHF 400 MHz
RG-174	Miniature	4 dB	20 dB
RG-58U	Small	2 dB	14 dB
RG-8X	Medium	1 dB	9 dB
RG-8U	Large	0.5 dB	4.5 dB
RG-213	Large	0.5 dB	4.5 dB
Hardline	Large, Rigid	nil	1.5 dB

Coax Cable Losses

T9C05 If you have a transmitter and an antenna which are 50 feet apart, but are connected by 200 feet of RG-58 coaxial cable, what should be done to reduce feed line loss?
  A. Cut off the excess cable so the feed line is an even number of wavelengths long.
  B. Cut off the excess cable so the feed line is an odd number of wavelengths long.
  C. Cut off the excess cable.
  D. Roll the excess cable into a coil which is as small as possible.

ANSWER C: Don't use any more cable than you must. Keep your cable lengths short.

T9C06 As the length of a feed line is changed, what happens to signal loss?
  A. Signal loss is the same for any length of feed line.
  B. Signal loss increases as length increases.
  C. Signal loss decreases as length increases.
  D. Signal loss is the least when the length is the same as the signal's wavelength.

ANSWER B: The longer your cable runs, the greater the line loss.

T9C07 As the frequency of a signal is changed, what happens to signal loss in a feed line?
A. Signal loss is the same for any frequency.
B. Signal loss increases with increasing frequency.
C. Signal loss increases with decreasing frequency.
D. Signal loss is the least when the signal's wavelength is the same as the feed line's length.
ANSWER B: In all feed lines, the higher the frequency, the higher the losses per 100 feet. This is why it's important to use only low-loss coax at VHF and UHF frequencies.

T9C08 If your antenna feed line gets hot when you are transmitting, what might this mean?
A. You should transmit using less power.
B. The conductors in the feed line are not insulated very well.
C. The feed line is too long.
D. The SWR may be too high, or the feed line loss may be high.
ANSWER D: If the feed line loss is too high, the line will rise in temperature because of the power dissipated in the line. This also will occur if there is a impedance mismatch to the antenna to produce a high standing-wave ratio on the feedline. Use low-loss coax for the feedline, and make sure you have a correct match to your antenna.

T9C09 Why should you make sure that no one can touch an open-wire feed line while you are transmitting with it?
A. Because contact might cause a short circuit and damage the transmitter
B. Because contact might break the feed line
C. Because contact might cause spurious emissions
D. Because high-voltage radio energy might burn the person
ANSWER D: Make sure no one can touch your open-wire feed lines. Both conductors have dangerous high voltages that will cause a nasty burn.

T9C10 For RF safety, what is the best thing to do with your transmitting antennas?
A. Use vertical polarization.
B. Use horizontal polarization.
C. Mount the antennas where no one can come near them.
D. Mount the antenna close to the ground.
ANSWER C: Since RF burns can be painful, make sure no one can accidentally touch your antenna.

T9C11 Why should you regularly clean, tighten and re-solder all antenna connectors?
A. To help keep their resistance at a minimum
B. To keep them looking nice
C. To keep them from getting stuck in place
D. To increase their capacitance
ANSWER A: A good antenna connection, finger tight at the back or top of your radio equipment, will keep resistance to a minimum, and will keep your signal strength at a maximum. To get all there is out of your two-way radio equipment, keep everything clean and tight, and have fun over Amateur Radio.

4

Novice Class Privileges

ABOUT THIS CHAPTER

Many of you learned the code as a Boy Scout or Girl Scout. Or maybe you were in the military service, and were required to learn the code. Ship captains and aircraft pilots also know the code. If you are one of these or someone that already knows the code, your first license step should be for the Novice Class license. This chapter describes Novice licensing and the privileges you will have when you satisfy the license requirements.

PAST NOVICE LICENSING

In 1951, the Novice Class operator's license was developed specifically to attract newcomers into the our hobby with a minimal code and theory test. Entry-level amateur operators were confined to crystal-controlled Morse code transceivers with 75 watts input power. The license was only good for one year, and not renewable.

In 1967, phase one of the FCC's new *incentive licensing* program approved a two-year term for Novices, and an applicant could re-apply for this same grade of license after a mandatory one-year wait after the two-year license expired. Power levels were increased, and variable frequency oscillators (VFOs) were allowed. Novices were even given a slice of the popular 2-meter band as an incentive to upgrade. Many increased their code speeds and went on to General Class.

NOVICE LICENSING TODAY

What started out as a temporary license to allow new ham students to get on the air with Morse code practice has now turned into a full-fledged amateur service Novice Class with very desirable, and permanent, privileges. Novice licenses are now good for 10 years, and may be renewed without any further testing.

In 1987, the Federal Communications Commission created *Novice Enhancement Voice Class*. This was one of the biggest incentives ever offered to beginners in the amateur service. Novices are now permitted to use long-range single-sideband to communicate by voice worldwide on 200 kHz in the choice 10-meter ham spectrum, which is regularly influenced by skywaves. Also, the world of VHF and UHF FM repeater operation has been opened to them on the 222 MHz band and on 1270

MHz. CW privileges are available for worldwide communications on 10 meters, 15 meters, 40 meters, and 80 meters.

CHANGING NOVICE TESTING PROCEDURES

The way a Novice applicant takes the test has changed over the years. When the Novice license was originally conceived, a single ham could administer the Novice test. A few years later, it took two hams, General Class or higher, and 18 years of age and older, to give the Novice test. It is changing again. Now the Novice testing is being folded into the very successful volunteer examination system.

Under the VEC System, it will take three accredited volunteer examiners (VEs) to administer the Novice test. These accredited VEs must hold a General Class, or higher, license and be at least 18 years of age, and have no business in the sale of Amateur Radio equipment or license preparation material.

Volunteer examination test sessions are everywhere. It will be easy for you to find a team of three VEs to give you a Novice written theory examination and a code test. Simply look in the back of this book for the list of volunteer examiner coordinators that is included and contact the coordinator nearest you. Or contact the ARRL or W5YI volunteer examiner coordinators who may offer their examinations anywhere in the world. Generally, one phone call will put you in touch with a VE team no more than 20 miles away! Examinations are usually given in the evenings, and on the weekends, all over the country. Your local Radio Shack store and Amateur Radio dealer are also good sources for the time and place of ham examination sessions.

THE WRITTEN THEORY EXAMINATION

You will study the first question pool in this book which covers all 350 questions in Element 2; 30 of these will be on your Novice written examination. Once you have passed the Novice written examination, you can concentrate on Element 3A, the Technician written examination. But why take both examinations together when you don't have to! For Novice Class, you need only to pass the Element 2 written examination. When you are ready to upgrade to Technician-Plus you have the Element 3A question pool right at your fingertips.

THE CODE TEST

For the Morse code requirement (Element 1A), you must pass the 5-wpm code test, which will give you credit for Element 1A. Morse code is an important part of your Amateur Radio study because you eventually may want to upgrade to General Class which requires knowing the code at 13 wpm. You will be required to receive code, but it is unusual for examiners to require you to send code.

Higher Code Speeds

Good news—you don't necessarily have to pass the slower code test to qualify for 5-wpm Element 1A credit. If you already know the code at a fast rate, skip the 5-wpm code test and try the Element 1B 13-wpm code test—or the Element 1C 20-wpm code test. Each test lasts approximately five minutes and is a typical CW transmission from one ham to another. Copy one minute of solid text, or score at least 70 percent on a 10-question test about the text, and you pass.

On Morse code examinations, you are permitted to skip over lower speed tests if you think you can pass the higher code rate. But this is only true on code tests—you must pass all the written examinations as you progress up the Amateur Radio ladder from Novice Class.

NOVICE CLASS PRIVILEGES

Passing the Novice examination gets you on the air BIG TIME. If you enjoy the Morse code, or if you like to talk worldwide, and on VHF and UHF frequencies, there is plenty of excitement ahead! Here is a brief description of each privilege.

80 METERS, 3675-3725 kHz

Novice operators are allowed to transmit up to 200 watts peak-envelope-power (PEP) output, using CW, for some exciting communications in code. Daylight range is approximately 100 miles, but nighttime range on 80 meters may span the continent with skywave propagation.

40 METERS, 7100-7150 kHz

By passing your code test and becoming a Novice operator, you may transmit CW up to 200 watts PEP output within this band for typical 300-mile daytime coverage, and 5000-mile nighttime skywave range. Also, 40 meters is a popular band for code practice stations. There's no telling what you can pick up to help increase your code speed.

15 METERS, 21,100-21,200 kHz

The 15-meter band normally enjoys worldwide skywave range during the day and early evening hours. Novice Class licensees may operate up to 200 watts PEP output CW to work the world. This is one of the best bands for extremely long range skywave contracts.

10 METERS, 28,100-28,500 kHz

Novice Class licensees may operate 200 watts PEP output CW for some very common worldwide and intercontinental contacts, thanks to skywave propagation. Skip signals on 10 meters are fun; they are best from May to September. Most CW activity is confined from 28,100 to 28,200 kHz. CW propagation beacons may be listened to between 28,200 and 28,300 kHz. Those propagation beacons let you know when the 10-meter band is "open" for skywaves.

10 METERS, 28,100-28,300 kHz

The Novice operator may also transmit digital privileges in this portion of the band. If you are into computers, you might transmit RTTY or packet communications near the bottom of your band. This is a great way to stay in touch with other hams who enjoy a computer hobby over the air.

10 METERS, 28,300-28,500 kHz

Novice Class operators may transmit single-sideband voice at 200 watts PEP output — plenty of power to work the world. Single-side-band (SSB) voice puts you right in the mainstream of worldwide activity when the 10-meter band is open. All ham operators with worldwide equipment love to exchange communications between 28,300 and 28,500 kHz because this is where all of the activity is located. This is your prime band for talking to the world over your microphone and high-frequency SSB transceiver.

1.25 METERS, 222.1-223.91 MHz

Novice licensees may operate up to 25 watts of PEP output on the 222-MHz band. All amateur modes and emissions are authorized to the Novice operator. Here is where you will find hundreds of repeaters throughout the country open for crystal-clear repeater and autopatch communications. A small handheld on the 222-MHz band will keep you in touch throughout the city. Handheld, mobile, and base equipment for the 222-MHz band is relatively inexpensive, and this is the ideal band to get started with your new Novice privileges using FM voice. Soon, your frequency privileges may be expanded, and you may also soon be allowed to be the control operator of a repeater system.

0.23 METERS, 1270-1295 MHz

This Amateur Radio band is located in the UHF spectrum, and is another popular one for line-of-sight repeater communications. The use of handheld and mobile equipment is plentiful. Antennas are extremely small at this UHF range. But since we are dealing with microwaves, power output to the Novice operator is limited to 5 watts PEP. This is the same power level that just about everyone uses on this band for safety purposes. Hundreds of FM repeaters are on the air, within your area, to extend your communications range. This band is also influenced by tropospheric ducting, and this lets you establish FM communications well beyond your local city area. Soon the frequency privileges may be expanded to give Novices the total band from 1240-1300 MHz.

NO 2-METER PRIVILEGES

What? No 2-meter privileges for Novice Class operators? That's right, the popular 2-meter band, 144-148 MHz, is reserved for Technician Class operators and higher. This is part of the incentive licensing program, and it will be an incentive to you, as a licensed Novice operator, to upgrade to Technician Class.

If you presently hold a Novice call sign and want to upgrade to Technician, you need only to pass an Element 3A written examination for instant access to the 2-meter band. Technically, your classification will be called Technician-Plus, meaning "plus" the code credit. Your "plus" credit will actually be a more powerful license than the typical Technician Class no-code license, because with Technician-Plus Class privileges, you may still continue to use 80 meters, 40 meters, 15 meters, and 10 meters. The Technician Class no-code operator doesn't receive privileges for these bands until he or she passes the 5-wpm code test.

GIVE IT A TRY—NOTHING TO LOSE

Let's say you take the Novice Class written examination and code test. You pass the theory, but don't pass the code. You still receive a credit good for one-year for passing the Element 2 theory examination! You may retry your code test at the same time, or at a later date up to 365 days away, and ultimately gain Novice privileges.

If you decide the code test is simply out of the question to pass at this time, then you may want to go for the Technician Class no-code license as your first license. If so, begin studying the Element 3A Technician question pool presented after the Element 2 question pool in this book. Maybe at some time in the future (no time limit), you may want to try the Element 1A 5-wpm code test again. If you pass it, you pick up the "plus" credit to become a Technician-Plus and gain more privileges.

We encourage you to join the amateur service through the Novice Class license. Learning the code is fun. After all, you will need to learn the code sometime if you ever want privileges on General Class, Advanced Class, and Extra Class frequencies. So give it a try—there is nothing to lose, and plenty to gain! Code is fun.

SUMMARY OF NOVICE OPERATING PRIVILEGES

Table 4-1 is a complete list of amateur operating privileges available to the Novice operator as of March 21, 1987, when Novice enhancement officially began. More good news is right around the corner—as mentioned previously, the Novice operators may soon gain additional frequencies in the 1.25-meter band, plus additional privileges in the 0.23 meter (23 centimeters) band.

Table 4-1. Novice Class Operating Privileges

Wavelength Band	Frequency	Emissions	Privileges
80 Meters	3675-3725 kHz	Code only	Limited to international Morse code. (200 watt PEP output limitation)
40 Meters	7100-7150 kHz	Code only	Same as above, except U.S. licensed operators in other than our hemisphere (ITU Region 2) are authorized 7050-7075 kHz due to shortwave broadcast interference. (200 watt PEP output)
15 Meters	21100-21200 kHz	Code only	Same as the 80-meter Novice privileges
10 Meters	28100-28500 kHz 28100-28300 kHz 28300-28500 kHz	Code Code Digital data Voice Code	Morse code. (200 watt PEP output) Digital and Morse code privileges. (200 watt PEP output.) Sideband voice and Morse code. (200 watt PEP output.)
1.25 Meters	222.1-223.91 MHz*	Code Digital data Voice Repeaters	All amateur modes and emissions authorized. This ham band does not exist except in our hemisphere. (25 watt PEP output transmitter power levels for Novice operators only. Technician-Plus and higher may operate at full amateur power levels.)
0.23 Meters	1270-1295 MHz*	Code Digital data Voice Repeaters TV	All amateur modes and emissions. Novice (only) transmitter power level limited to 5 watts PEP transmitter output.

*The FCC has proposed to allow Novice operation between 222-225 MHz, and to permit Novice Class repeater control operators at 1.25 meters and 0.23 meters.

5

Learning Morse Code

ABOUT THIS CHAPTER

If you decide to enter the amateur radio service via the Novice Class license, you will need to demonstrate proficiency in 5-wpm Morse code telegraphy. This chapter tells you how. It tells you how to use the *Learning Morse Code* training tapes that are available at Radio Shack stores and other ham radio dealers, and what to expect when you take the code test.

LOOKING AT MORSE CODE

The International Morse code, originally developed as the American Morse code by Samuel Morse, is truly international—all countries use it, and most commercial worldwide services employ operators that can recognize it. It is made up of short and long duration sounds. Long sounds, called "dahs," are three times as long as short sounds, called "dits." *Figure 5-1* shows the time intervals for Morse code sounds and spaces. *Figure 5-2* indicates the sounds for the characters and symbols.

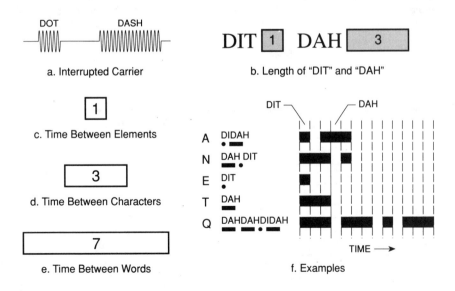

Figure 5-1. Time Intervals for Morse Code

LETTER	Composed of:	Sounds like:	LETTER	Composed of:	Sounds like:
A	• —	didah	N	— •	dahdit
B	— • • •	dahdididit	O	— — —	dahdahdah
C	— • — •	dahdidahdit	P	• — — •	didahdahdit
D	— • •	dahdidit	Q	— — • —	dahdahdidah
E	•	dit	R	• — •	didahdit
F	• • — •	dididahdit	S	• • •	dididit
G	— — •	dahdahdit	T	—	dah
H	• • • •	didididit	U	• • —	dididah
I	• •	didit	V	• • • —	didididah
J	• — — —	didahdahdah	W	• — —	didahdah
K	— • —	dahdidah	X	— • • —	dahdididah
L	• — • •	didahdidit	Y	— • — —	dahdidahdah
M	— —	dahdah	Z	— — • •	dahdahdidit

a. Alphabet

CHARACTER		Composed of:	Sounds like:
AR	(end of message)	• — • — •	didahdidahdit
K	invitation to transmit (go ahead)	— • —	dahdidah
SK	End of work	• • • — • —	dididahdidah
S O S	International distress call	• • • — — — • • •	dididahdahdahdididit
V	Test letter (V)	• • • —	didididah
R	Received, OK	• — •	didahdit
BT	Break or Pause	— • • • —	dahdidididah
DN	Slant Bar	— • • — •	dahdidididahdit
KN	Back to You Only	— • — — •	dahdidahdahdit
Period		• — • — • —	didahdidahdidah
Comma		— — • • — —	dahdahdididahdah
Question mark		• • — — • •	dididahdahdidit

b. Special Signals and Punctuation

NUMBER	Composed of:	Sounds like:
1	• — — — —	didahdahdahdah
2	• • — — —	dididahdahdah
3	• • • — —	didididahdah
4	• • • • —	dididididah
5	• • • • •	dididididit
6	— • • • •	dahdidididit
7	— — • • •	dahdahdididit
8	— — — • •	dahdahdahdidit
9	— — — — •	dahdahdahdahdit
0	— — — — —	dahdahdahdahdah

c. Numerals

Figure 5-2. Morse Code and Its Sound

CODE REQUIREMENTS FOR AMATEUR RADIO CLASSES

Table 5-1 shows again the code requirements for the Novice, General, and Extra Class licenses.

Table 5-1. Amateur Radio Classes Code Requirements

Novice Voice Class license	5 wpm, sending and receiving plain language text
General Class license	13 wpm, receiving plain language text
Extra Class license	20 wpm, receiving plain language text

Commercial operators must pass a much tougher code examination—in addition to plain language, their test also consists of five-letter word groups that must be copied perfectly for one minute out of five. For the ham radio tests, you don't need perfect copy. There are many acceptable answer formats; which one is used is determined by your VE team.

WHY CODE?

Learning the code to obtain an amateur operator/primary station license is an old tradition. When you pass the Element 1A 5-wpm code test for the Novice Class license, you immediately gain worldwide voice privileges on 10 meters, and CW worldwide privileges on 10 meters, 15 meters, 40 meters, and 80 meters.

You will need code tapes or a CW computer program to get started learning the code. These are available at Radio Shack stores and throughout the Amateur Radio marketplace. Obtain as many code training materials as possible where you learn by listening to the sounds of dits and dahs. This way, code will be fun, and you will easily pass the 5-wpm code test.

CODE KEY

All Morse code is sent by using a code key. A typical one is shown in *Figure 5-3a*. Normally it is mounted on a thin piece of wood or plexiglass. You can use wood screws, or simply glue the key in place. Make sure that what you mount it on is *thin*; if the key is raised too high, it will be uncomfortable to the wrist. The correct sending position for the hand is shown in *Figure 5-3b*.

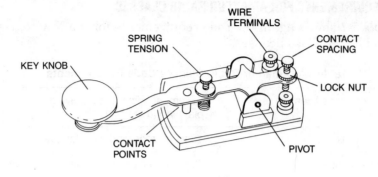

a. Code Key

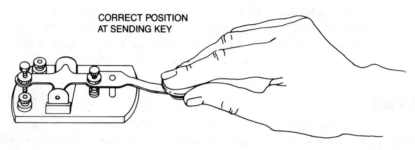

b. Sending Position

Figure 5-3. Code Key for Sending Code

LEARNING MORSE CODE

The main reason you are learning Morse code is to be able to pass an Element 1A 5-wpm code test which is part of your entry-level Novice Class examination (FCC Rule 97-503). Here are some suggested ways for you to learn Morse code:

1. Use cassette tapes available from Radio Shack (62-2418) or ham radio dealers.
2. In addition to the cassette tapes, practice by tuning into live code broadcasts with a ham radio or shortwave receiver, or attend a code class.
3. Use a code key and oscillator for practice by yourself and with a friend.
4. Use learning Morse code computer software programs.

Let's first consider using the cassette tapes.

Cassette Tapes

Five words per minute is so slow, and so easy, that many ham radio applicants learn it completely in a single week! You can do it too by using the code tapes mentioned above.

Code cassettes personally recorded by your author make code learning *fun*. They will train you to send and receive the International

Morse code in just a few short weeks. They are narrated and parallel the instructions in this book. The available cassette tapes mentioned have code characters generated at a 15-wpm character rate, spaced out to a 5-wpm word rate. This is known as Farnsworth spacing.

Getting Started

THE HARDEST PART OF LEARNING THE CODE IS TAKING THE FIRST CASSETTE OUT OF THE CASE, PUTTING IT IN YOUR CASSETTE PLAYER, AND PUSHING THE PLAY BUTTON! Try it, and you will be over your biggest hurdle. After that, the tapes will talk you through it in no time at all.

The cassette tapes make code learning *fun*. You'll hear how humor has been added to the learning process to keep your interest high. Since ham radio is a hobby, there's no reason we can't poke ourselves in the ribs and have a little fun learning the code as part of this hobby experience. Okay, you're still not convinced—you probably have already made up your mind that trying to learn the code will be the hardest part of getting your Novice Class license. It will not. Give yourself a fair chance. Don't get discouraged. Have patience and remember these important reminders when practicing to learn the Morse code:

■ Learn the code by sound. Don't stare at the tiny dots and dashes that we have here in the book—the dit and dah sounds on the cassette and on the air and with your practice keyer will ultimately create an instant letter at your fingertips and into the pencil.

■ *Never* scribble down dots or dashes if you forget a letter. Just put a small dash on your paper for a missed letter. You can go back and figure out what the word is by the letters you did copy!

■ Practice only with fast code characters; 15-wpm character speed, spaced down to 5-wpm Novice Class speed is ideal. The FCC recommendations are quite clear about the rate that code characters are generated for each class of code examination. You will find the code speed, character speed, and actual tone (1000 Hz) on cassette courses identical to cassettes used by most hams that give code tests.

■ Practice the code by writing it down whenever possible. This further trains your brain and hand to work together in a subconscious response to the sounds you hear (Remember Pavlov and his dog "Spot"?)

■ Practice only for 15 minutes at a time. The tapes will tell you when to start and when to stop. Your brain and hand will lose that sharp edge once you go beyond 16 minutes of continuous code copy. You will learn much faster with five 15-minute practices per day than a one-hour marathon at night.

■ Stay on course with the cassette instructions. Learn the letters, numbers, punctuation marks, and operating signals in the order they are presented here. Your author's code teaching system parallels that of the American Radio Relay League, Boy Scouts of America, the Armed Forces, and has worked for thousands in actual classroom instruction.

It was no accident that Samuel Morse gave the single dit for the letter "E" which occurs most often in the English language. He determined the most used letters in the alphabet by counting letters in a printer's type case. He reasoned a printer would have more of the more commonly used letters. It worked! With just the first lesson, you will be creating simple words and simple sentences with no previous background.

Table 5-2 shows the sequence of letters, punctuation marks, operating signals, and numbers covered in six lessons cn the cassettes recorded by your author.

Table 5-2. Sequence of Lessons on Cassettes

■ Lesson 1	E T M A N I S O $\overline{SK}$ Period
■ Lesson 2	R U D C 5 0 $\overline{AR}$ Question Mark
■ Lesson 3	K P B G W F H $\overline{BT}$ Comma
■ Lesson 4	Q L Y J X V Z $\overline{DN}$ 1 2 3 4 6 7 8 9
■ Lesson 5	Random code with narrated answers
■ Lesson 6	A typical Novice Class code test

Code Key and Oscillator—Ham Receiver

All worldwide ham transceivers have provisions for a code key to be plugged in for both CW practice off the air, as well as CW operating on the air. If you already own a worldwide set, chances are all you will need is a code key for some additional code-sending practice.

Read over your worldwide radio instruction manual where it talks about hooking up the code key. For code practice, read the notes about operating with a "side tone" but not actually going on the air. This "side tone" capability of most worldwide radios will eliminate your need for a separate code oscillator.

Code Key and Oscillator—Separate Unit

Many students may wish to simply buy a complete code key and oscillator set. They are available from local electronic outlets or through advertisements in the ham magazines.

Look again at the code key in *Figure 5-3a*. Note the terminals for the wires. Connect wires to these terminals and tighten the terminals so the wires won't come loose. The two wires will go either to a code oscillator set or to a plug that connects into your ham transceiver. Hook up the wires to the plug as described in your ham transceiver instruction book or the code oscillator set instruction book.

Mount the key firmly as previously described, then adjust the gap between the contact points. With most new telegraph keys, you will need a pair of pliers to loosen the contact adjustment knob. It's located on the very end of your keyer. First loosen the lock nut, then screw down the adjustment until you get a gap no wider than the thickness of a business card. You want as little space as possible between the points. The points are located close to the sending plastic knob.

Now turn on your set or oscillator and listen. If your hear a constant tone, check that the right-hand movable shorting bar is not closed. If it is, swing it open. Adjust the spring tension adjustment screw so that you get a good "feel" each time you push down on the key knob. Adjust it tight enough to keep the contacts from closing while your fingers are resting on the key knob.

PICK UP THE KEY BY THE KNOB! This is the exact position your fingers should grasp the knob—one or two on top, and one or two on the side of it. Poking at the knob with one finger is unacceptable. Letting your fingers fly off the knob between dots and dashes (dits and dahs) is also not correct. As you are sending, you should be able to instantly pick up the whole key assembly to verify proper finger position.

Your arm and wrist should barely move as you send CW. All the action is in your hand—and it should be almost effortless. Give it a try, and look at *Figure 5-3b* again to double-check your hand position.

Letting someone else use the key to send to you will also help you learn the code.

Morse Code Computer Software

The newest way to learn Morse code is through computer-aided-instruction. There are many good PC programs on the market that not only teach you the characters, but build speed and allow you to take actual telegraphy examinations which the computer constructs. Personal computer programs are also often used to make audio tapes on your tape recorder so you can listen to them on the cassette player in your car.

A big advantage of computer-aided Morse code learning is that you can easily customize the program to fit your own needs! You can select the sending speed, Farnsworth character-spacing speed, duration of transmission, number of characters in a random group, tone frequency—and more!

Some have build-in "weighting." That means the software will determine your weaknesses and automatically adjust future sending to give you more study on your problem characters! All Morse code software programs transmit the tone by keying the PC's internal speaker.

THE ACTUAL CODE TEST

Almost all Morse code examinations for Element 1A, 5 wpm; Element 1B, 13 wpm; and Element 1C, 20 wpm are plain language communications from one ham to another. *Figure 5-4* shows an example of a Novice Class 5-wpm code test.

5 wpm

VVV VVV NI3R DE N5CRH BT RRR AND TNX SUZY. THE ANTENNA IS UP JUST 28 FEET AND IS A HUSTLER RM10 ON MY REAR BUMPER. BY THE WAY, I STILL HAVE PROBLEMS COPYING, / 467 9 KQ. HOW COPY NOW? NI3R DE N5CRH AR SK

Figure 5-4. A Typical Novice Code Test

On the Novice 5-wpm CW test, the actual character rate will be generated at approximately 15 wpm—18 wpm with a pitch of 1000 Hz. The W5YI VE system, plus many other VE systems, use a 15-wpm character rate with big spaces in between for a 5-wpm word rate. This is called Farnsworth code.

The American Radio Relay League (ARRL) uses a snappier 18-wpm character rate, with slightly longer spaces to slow the word rate down to 5 wpm. This is known as fast Farnsworth.

You should listen to code preparation tapes featuring both character rates and pick the VE system that has the character rate that you like the most. Some like the faster snappier 18-wpm character rate and others may appreciate a seemingly slower sound code at 15-wpm character rate. Even though these two character rates differ, the spaces in between the letters bring the code speed down to an actual 5-wpm word rate.

For the Novice code test, the examiner can choose to ask you 10 questions about the message you copied. If you answer correctly, you pass. However, in many cases, it's difficult for examiners to ask 10 questions about the copy you have written. Rather, most examiners will look over your copy for *one minute of perfect copy*. All you need are five consecutive words and you have it! Usually, after the code test is finished, they will give you a few minutes to go back and clean up your code copy.

Preceding the test will be a one-minute warm-up. This lets you get used to the sound of their code test equipment, and allows you to tell the examiners whether or not you want the volume raised or lowered. Some examination teams can change the character rate to suit you. Ask about it ahead of time so the VET can give you the best chance to pass.

Once everyone is ready for the test to begin, the examiners start the equipment. Most CW code tests start with a series of Vs. Then get set for call signs. After the call signs, you will be writing down a plain language text that will contain every letter, every number, and the punctuation marks required by the FCC rules. Numbers, punctuation marks and operating procedure signs count as two characters on the test.

That's right, every letter, every number, and all required punctuation marks must be used during the approximate 5-minute code test for Novice. This means you might expect those harder letters to appear in the call signs. If you miss a call sign, keep right on copying—you'll have more than enough opportunity to pick up five consecutive words when they talk about the weather, type of antenna, type of rig, QTH, or other common elements within the usual on-air QSO. Refer again to *Figure 5-4* for the 5-wpm sample test and study it to get a feel for the flow of the message.

If you listen to the author's code test preparation tapes for Novice, you will pick up some additional skills on breezing through the Novice 5-wpm code test. You *should* listen to outside code testing materials to *hear* the sound of Morse code, rather than just looking at it in this book. While you might be able to memorize the code by reading over this book, you *must* take the code test by ear to pass it. So get those code tapes, and a computer program, and get started listening to the code.

Morse code is fun, and you will ultimately need to obtain the Morse code elements as you upgrade to higher operating privileges within the amateur service. So why not get started—RIGHT NOW!

Taking the Test

To take the actual test, your examiners will have a separate setup for your code test, different from the written examination. The examiners will send a typical transmission from one ham to another. This will consist of an exchange of call signs, names, signal reports, the local weather, the type of rig or antenna the operators are using, occupation, and anything else that normally takes place in a ham radio conversation. You write down as much as you can, without worrying about missing a letter here or there.

After the Test

After the test is over, you can go back and work on your copy to fill in where you missed a letter. Since they were using plain language, it's easy to spot letters you accidentally copied wrong, or misspellings because you accidentally wrote an "A" instead of an "N".

When you think you have done the best in correcting your material, the examiners will look for one minute of perfect copy, or may ask you ten simple questions about what you copied. These questions might have fill-in-the-blank answers, or better yet, multiple-choice answers.

Seven out of ten is usually a passing score. They might also send another code transmission if you need a second chance—there is no longer a required waiting time between missed examination elements.

SENDING CODE

Some examiners might ask you to use a straight key to send a few words in code to them. The rules (97.503a) are quite clear that code sending is required, but many examiners don't ask you to send—only to receive. The FCC does not require a sending test since it has been their experience that receiving proficiency usually indicates you can send code as well. We hope that you will send with your own code practice setup to show the examiners how well you know the code and theory course! Sending also helps you practice pleasant sounding CW.

Novice and Beyond

Even if you enter the amateur service with the Technician Class no-code license, you may get the Morse code credit at a later time by taking the Element 1A code test. When you pass this test, you will instantly become a Technician-Plus Class operator.

Later on when you take the General Class code test (Element 1B), it will be administered the same way. For the General Class, the code speed is 13 wpm, with the actual characters sent at 15-18 wpm. No sending test is required by the FCC, although your examiners have the option of requiring it.

For the Extra Class license, the very highest ham radio license available, the code requirement is 20 wpm, usually sent at 21 wpm. Just like all other examinations, one minute of perfect copy or seven out of ten questions answered correctly passes the code test (Element 1C).

NOW WE'RE READY

Obtain the code practice materials that will best suit your listening pleasure. If you regularly work with computers, we've mentioned the excellent computer programs for learning code. If you regularly listen to tape cassettes in your automobile, we've mentioned the code learning cassettes developed and recorded by your author. Both of these products are available at Radio Shack and in the Amateur Radio marketplace.

You might also consider purchasing your high frequency equipment ahead of time, in preparation for your privileges on 10 meters, 15 meters, 40 meters, and 80 meters. You don't need a license to buy ham equipment; however, don't *transmit* until your license arrives from the FCC. *Listen* to the code practice on 80 and 40 meters at night, and at 15 and 10 meters during the day. This is another great way to prepare yourself for the Novice 5-wpm code test.

6

Technician-Plus Class Privileges

ABOUT THIS CHAPTER

A step up the amateur service ladder from the entry-level licenses of Technician Class no-code or Novice Class is the Technician-Plus Class. Ham operators who have entered the amateur service without code by passing the Technician Class no-code license requirements need only pass an Element 1A 5-wpm code test to be awarded a Technician-Plus Class upgrade. With the code proficiency, they gain the long-distance DX HF privileges on the 10-, 15-, 40-, and 80-meter bands.

Amateur operators that already have demonstrated code proficiency by passing the Novice Class requirements need only pass a 25-question written examination on Element 3A and they will have earned Technician-Plus upgrade privileges.

This chapter details the operating privileges available to a Technician-Plus Class operator. They end up being the combined privileges of the Novice Class license and the Technician Class no-code license.

TECHNICIAN-PLUS CLASS OPERATING PRIVILEGES

Table 6-1 highlights the added privileges you will receive on 80 meters, 40 meters, 15 meters, and 10 meters by passing the Technician-Plus Class requirements.

80-METER WAVELENGTH BAND, 3500-4000 kHz

Novice and Technician-Plus Class privileges on the 80-meter band are Morse code only from 3675 to 3725 kHz. These frequencies have recently been shifted to accommodate our Canadian neighbor's band plan. A Technician Class no-code operator needs to pass at least a 5-wpm code test to get onto this band for CW work.

40-METER WAVELENGTH BAND, 7000-7300 kHz

Passing a 5-wpm code test gives you Morse-code-only privileges on this band at 7100 to 7150 kHz. This is a popular Novice Class and Technician-Plus Class CW band because evening QSO's can reach up to 5000 miles away.

Table 6-1. Technician-Plus Class Operating Privileges

Wavelength Band	Frequency	Emissions	Comments
160 Meters	1800–2000 kHz	None	No privileges
80 Meters	3675–3725 kHz	Code only	**Limited to Morse code (200 watt PEP output limitation)
40 Meters	*7100–7150 kHz	Code only	**Limited to Morse code (200 watt PEP output limitation)
30 Meters	10100–10150 kHz	None	No privileges
20 Meters	14000–14350 kHz	None	No privileges
17 Meters	18068–18168 kHz	None	No privileges
15 Meters	21100–21200 kHz	Code only	**Limited to Morse code (200 watt PEP output limitation)
12 Meters	24890–24990 kHz	None	No privileges
10 Meters	28100–28500 kHz 28100–28300 kHz 28300–28500 kHz	Code Data and code Phone and code	Morse code (200 watt PEP output code limitation) Sideband voice (200 watt code PEP output limitation)
6 Meters	50.0–54.0 MHz	All modes	Morse code, sideband voice, radio control, FM repeater, digital computer, remote bases, and autopatches (1500 watts PEP output)
2 Meters	144–148 MHz	All modes	All types of operation including satellite and owning repeater and remote bases. (1500 watt PEP output)
1¼ Meters	220–225 MHz	All modes	All band privileges. (1500 watt PEP output)
70 cm	420–450 MHz	All modes	All band privileges, including amateur television, packet, RTTY, FAX, and FM voice repeaters. (1500 watt PEP output)
35 cm	902–928 MHz	All modes	All band privileges. Plenty of room! (1500 watt PEP output.)
23 cm	1240–1300 MHz	All modes	All band privileges. (1500 watt PEP output)

* U.S. licensed operators in other than our hemisphere (ITU Region 2) are authorized 7050-7075 kHz due to shortwave broadcast interference.
** No change from Novice Privileges.

15-METER WAVELENGTH BAND, 21,000-21,450 kHz

Technician-Plus Class via the 5-wpm code test provides you CW privileges from 21,100 to 21,200 kHz in this portion of the worldwide band. Novice Class and Technician-Plus Class operators can expect ranges to distant stations in excess of 10,000 miles during daylight hours. Nighttime DX activity on this band normally fades away about 9:00 P.M. local time.

10-METER WAVELENGTH BAND, 28,000-29,700 kHz

Passing a code test at 5 wpm allows you to operate on this very popular band for regular worldwide privileges. Novice Class and Technician-Plus Class operators may operate code and digital computer privileges from 28,100 to 28,300 kHz, and monitor 28,200 to 28,300 kHz for low-power propagation beacons. Novice Class and Technician-Plus Class operators, having passed a code test, are allowed code and voice privileges between 28,300 and 28,500 kHz. It's worth it to learn the code for this popular band after you have successfully passed your Technician Class license examination.

EXAMINATION ADMINISTRATION

Up-grades to Technician-Plus are administered the same as the Technician Class no-code examination. Three accredited VEs will form the examination team. If you have a Novice Class license, they will administer the 25-question written theory examination on Element 3A. You must get 74 percent of the questions (22) correct on the written examination. If you are a Technician Class no-code operator, they will administer the Element 1A 5-wpm code test.

The code test is described in Chapter 5, *Learning Morse Code.* Refer to it for aid in learning code, what the actual test will be like, and how the examiners determine if you pass.

Remember, you may take another test immediately if you fail either the theory examination or the code test. And another reminder, if you have a Technician Class no-code license, you will not receive another Technician license. Your CSCE validates your Technician-Plus operating privileges.

HIGHER CODE SPEEDS

As with Novice, you don't necessarily have to pass the slower code test to qualify for the Element 1A 5-wpm credit. If you already know code at a fast rate, ask for an Element 1B 13-wpm code test or an Element 1C 20-wpm code test. If you copy one minute of solid text, or score 70 percent on a 10-question test about the copied text, then you pass and receive credit for the higher rate.

THE NEXT STEP

Once you have met the requirements for your upgrade to Technician-Plus why not keep going. General Class requires a 25-question written theory examination on Element 3B question pool and an Element 1B 13-wpm code test. Why not get started, build your code speed, study the Element 3B question pool, and go for it! Radio Shack and ham radio dealers have your author's book *General Class* and tapes to make the upgrade an easy one.

7

Taking the License Examinations

ABOUT THIS CHAPTER

If you are preparing for your Novice Class examination, you will have studied Element 2 out of this book, plus you will have practiced your code using audio cassettes or computer code programs.

If you are preparing for your Technician Class no-code license, you will have studied both Elements 2 and 3A question pools, and you are ready for both examinations. No code test is required for the no-code Technician, but try it anyway — you might end up a Technician-Plus operator.

This chapter tells you how the examinations will be given, who is qualified to give them, and what happens after you complete them.

EXAMINATION ADMINISTRATION

For Technician Class No-Code License

The FCC no longer conducts Amateur Radio service examinations; all examinations are conducted by volunteer amateur operators. The test sessions are coordinated by national or regional volunteer-examiner coordinators (VECs) who accredit General Class, Advanced Class and Extra Class ham operators to serve as volunteer examiners (VEs).

Three officially certified VEs form a volunteer examination team (VET) which is required to administer a Technician Class examination. The VEs are not compensated for their time and skills, but they are permitted to charge you a fee for certain reimbursable expenses incurred in preparing, processing, or administering the examination. The maximum fee is adjusted annually by the FCC. It is less than $6.00.

Now that the Novice Class examination is being folded into the VEC system, you may take one or both examinations in front of the same VET. Contact your VET ahead of time and let them know you are planning to attend one of their upcoming test sessions. In addition, though Novices paid no fee previously, the FCC will probably allow the VE fee for Novice as well.

The VETs offer local Technician Class examinations regularly at local sites to serve their community. They generally coordinate closely and rotate examination sites so you should be able to find a test site near you. You can obtain information about VETs and examination sessions by checking with your local radio club, ham radio store, and local packet bulletin boards. Also, the local Radio Shack manager can often prove helpful here. If this is not convenient, write or call the VEC that serves your area. The VEC will be able to give you the name, address and telephone number of your local VET. A list of VECs that was current at the time of publication is given in the Appendix.

Once you have found your examination location and set the possible date for the test session, contact the local VET examiners and pre-register for taking your examination. They will hold a seat for you at the next available session, and they will appreciate you contacting them ahead of time making a reservation. Don't be a no-show, and don't be a surprise-show. Call them ahead of time!

For Technician-Plus Class Upgrade

With Technician No-Code License

Your examination will be administered just as for the Technician Class no-code license except that you will only be signing up for an Element 1A 5-wpm Morse code test.

With Novice License

Your examination will be administered just as for the Technician Class no-code license except that you will only be signing up for an Element 3A written examination.

With a Knowledge of Code and No Operator License

Your examination will be administered just as for the Technician Class no-code license except that you will be signing up for the Element 2 written examination, the Element 3A written examination, and the Element 1A 5-wpm code test. All three may be taken in one sitting.

Special Cases

You may have chosen to start with the Novice Class and have passed the Element 2 written examination, but have not passed the Element 1A 5-wpm code test. If it has not been 365 days since you passed the Element 2 written examination, you should sign up for a Element 3A written examination with the VET. Your administration will be the same as for a Technician Class no-code license. Bring your Form 610 and/or CSCE to the test session to assure that you get credit for passing the Element 2 written examination. You will be awarded a Technician Class no-code operator's license when you pass the Element 3A written examination.

If you have passed the Element 1A code test and a year has not passed, but have not passed the Element 2 or Element 3A written examinations, you should sign up for the two written examinations. Your examinations will be administered just as for the no-code Technician Class, but you will be awarded Technician-Plus Class operator's privileges if you pass both examinations.

EXAM CONTENT

The FCC previously handled amateur services testing. They developed the questions, the multiple choices, and identified the correct answers. And it was all sort of secret! Neither the questions, nor the correct answers, were really widely known. That has changed. Since 1982, when President Reagan signed legislation providing for volunteer amateur operator examinations above the Novice Class, the FCC has been transferring, in phases, testing responsibility—including development of examination questions—to the amateur service community. VECs periodically revise the questions in the various amateur operator class question pools, and recommend multiple-choice answers to the VEs. The VEs are responsible for the answers, but most VEs accept the multiple-choice answers, both correct and incorrect, supplied by the VECs.

What most VECs and VEs have adopted is an amateur services examination system similar to that used by the FAA for testing pilots. Pilots know all the possible questions that might be on an examination. Amateur operators are given that same privilege with their questions when they take an examination. If it works for the FAA, it should work for the FCC!

COMPLETING THE FCC FORM 610

An FCC Form 610, *Application for Amateur Radio Station / Operator License*, is included in the back of this book. Carefully tear out this form and bring it to the examination site.

When you call your volunteer examiner to confirm your attendance at a testing session, ask them whether or not you should begin to fill out your portion of Section I of FCC Form 610. If they tell you to go ahead and fill out as much as you can ahead of time, answer Questions 2C, 5, 6, 7, 8, 9, 10. Sign on Line 13 and put the date in Box 14. Don't make any marks on the top of the form, nor do anything on the back of the form — this is for your volunteer examiners. Be sure your signature on Line 13 matches your name on Line 5, and make sure they can read your printing. Use the example in the Appendix as a guide.

Item 7 is your current mailing address. Where do you wish to receive your license? Where do you normally get your mail? This may be any postal address that will get your license to you when it is issued.

Since you are allowed to operate your amateur station anywhere in the United States, Item 8 (Current Station Location) used to be a physical address where the FCC might go to track you down. It was, in many cases, different from Item 7 because it could not be a P.O. box number or a rural route number. However, effective March 1, 1993 FCC amateur service licenses will no longer indicate a station address which is different from the licensee's mailing address. It is no longer necessary to complete Item 8 on the Form 610 application. Future versions of the form will eliminate 2H (Change Station Location) and Item 8. Even so, the Form 610 included in this book may be used indefinitely, and even if information is entered in 2H and Item 8, the FCC will ignore it.

Take a look at Item 6. Did you get your birthdate down correctly? Make sure the last two numbers aren't this year's date! Items 9 and 10 are normally answered "No" and "No". Check the instructions to make sure this is correct for you. Item 11 and Item 12 are normally left blank unless you are upgrading and have not yet received your previous license from the FCC.

Your Examiner's Portion

The VET will complete the administering VE's report on the front, certify the back of the FCC Form 610, and send it in to their VEC for processing. For new applications, it takes approximately 6 to 8 weeks to finally receive your official FCC call sign. Your call letters will be sent directly to the mailing address you entered on your Form 610.

SPECIAL PROVISIONS FOR THE HANDICAPPED

The amateur service welcomes handicapped operators. The FCC rules encourage examiners to take any necessary steps to assist the handicapped operator through the examination process. For the visually impaired, the examiners could read you the questions and ask for spoken answers. If you're not able to write clearly, the examiners will assist you in transcribing the answers onto the answer sheet. As of February, 1991, the FCC will allow exemptions from the 13- and 20-wpm code tests for amateur operator licensees who are incapable of passing the higher speed code tests due to severe handicaps. If you are handicapped, you are encouraged to document your physical handicap with correspondence from your physician so that your VET will know how they might assist you in getting through the testing process. This procedure protects the VEs from unwarranted criticism of their judgement to extend special privileges. A physician's certificate of disability, which can be used to support an exemption of the 13- or 20-wpm code test, is part of the FCC gold Form 610 in the Appendix.

Be sure to tell the VET ahead of time that you may require a special examination for the handicapped—as long as they know that you are coming with special circumstances, they will work closely with you to help you pass the test. Handicapped operators are a valuable resource of quality trained ham radio operators.

TAKING THE EXAMINATION

Get a good night's sleep before exam day. Continue to study both theory elements up to the moment you go into the room. Though not presently available at Radio Shack, other amateur radio dealers may have Gordon West cassette tapes that cover the theory needed for the

Element 2 and Element 3A written examinations. You may want to purchase a set and listen to them while you are waiting to take the test. Listen to the *Learning Morse code* tapes available from Radio Shack in your car as you drive to the examination session if you are going to take a code test.

What to Bring to the Exam

Here's what you'll need for your Novice Class or Technician Class written examinations or optional code test:

1. Examination fee of approximately $6.00 in cash.
2. Personal identification with a photo.
3. Properly filled out FCC Form 610.
4. Any other Certificates of Successful Completion of Examinations (CSCEs) or signed 610 forms from other examiners who have tested you within 365 days of this test session date. Bring the originals, plus two copies of everything.
5. Some sharp pencils and fine-tip pens. Bring a backup!
6. Calculators may be used, so bring your calculator.
7. A letter from your physician indicating you are handicapped if you are requesting a special handicapped exam. If you need any special equipment, such as a braille writer, you must supply it.
8. Any other item that the VET asks you to bring. Donuts are always welcome. Remember, these volunteer examiners receive no pay for their work.

Check and Double-Check

Read over the examination questions carefully. Take your time in looking for the correct answer. Some answers start out looking correct, but end up wrong. Don't speed read the test.

When you are finished with an examination, go back over every question and double-check your answers. Try a game where you read what you have selected as the correct answer, and see if it agrees with the question.

When the examiners hand out the examination material, put your name, date, and test number on the answer sheet. *Make No Marks On the Multiple-Choice Question Sheet.* Only write on the answer sheets.

When you are finished with the examination, turn in all of your paperwork. Tell the examiners how much you appreciate their unselfish efforts to help promote ham radio participation. If you are the last one in the room, volunteer to help them take down the testing location. They will appreciate your offer.

And now wait patiently outside for the examiners to announce you have passed both examinations. Chances are they will greet you with a smile and your certificate of completion. Make sure to immediately sign this certificate of completion when it is handed to you.

YOUR NEXT STEP

Did you try the code test? If not, your next step after no-code Technician is Technician-Plus. This only requires the 5-wpm code test. If you haven't purchased cassette code tapes to learn code and build your speed, it is suggested that you do so.

It is recommended that you also begin thinking about General Class. If you decide to learn code, go ahead and shoot for the stars, and prepare for passing the code test at General Class speed — 13-wpm. Radio Shack has license preparation materials for General Class. There is a Gordon West authored book, *General Class*, with the Element 3B question pool to make it a breeze to pass the 25-question General Class written examination. Also, there are code cassettes that build code speeds from 5 wpm to 13 wpm to help you pass the Element 1B 13-wpm code test.

So don't stop now at Technician or Technician-Plus—go for General Class! That's your next step.

TECHNICIAN CLASS CALL SIGNS

Technician Class call signs are selected from category Group C by a computer in Gettysburg, Pennsylvania. It's a sequential selection, so sorry, you can't pick out a favorite. Your call letters will consist of a single letter, a single number, followed by 3 letters (N6NOA). Unfortunately, there are some parts of the country where all of the Group C call letters have been used up. This means you may receive a Group D set of Technician call letters, beginning with 2 letters, a number, and followed by 3 letters (KB1XYZ).

But whatever call sign you get, you will be proud of it because you are the only one in the world with that particular FCC-issued call sign.

When you upgrade to Advanced Class, you have another opportunity to change call signs.

CONGRATULATIONS! YOU PASSED

After you pass the examinations, congratulations are in order and a big welcome to Technician Class privileges. The world of microwave and VHF/UHF operating awaits you. And, if you also passed the code test, welcome to Technician-Plus privileges and the world of long-range high-frequency operation.

You cannot begin operating until your official FCC call sign arrives. It will take from 6 to 8 weeks for your call sign to be processed and sent to the address listed on your Form 610. You can buy equipment ahead of time, but hide the microphone until your ticket arrives in the mail. You cannot go on the air until your own call sign hits town.

SUMMARY

Welcome to the world of Amateur Radio. Maybe I'll hear you on the airwaves soon. Here in Southern California, I'm on the 2-meter band on 144.330 MHz. If you have Technician-Plus privileges, you can find me on 10 meters at 28.303 most mornings. And if the band is open, I'm also on the worldwide 6-meter band near 50.120 MHz.

I would also like you to write me and send us a triple-stamped, self-addressed envelope for an exclusive Novice or Technician Class passing certificate. I'll sign it personally, and I welcome you to the fascinating world of Amateur Radio. My address is in the Appendix.

Hope to hear you on the bands soon.

Gordon West, WB6NOA

Appendix

U.S. VOLUNTEER-EXAMINER COORDINATORS IN THE AMATEUR SERVICE

Anchorage Amateur Radio Club
2628 Turnagain Parkway
Anchorage, AK 99517
(907) 786-8121 (Day)
(907) 243-2221 (Night)
(907) 276-5121
(907) 274-5546

ARRL/VEC*
225 Main Street
Newington, CT 06111-1492
(203) 666-1541
FAX (208) 665-7531

Central Alabama VEC, Inc.
1215 Dale Drive SE
Huntsville, AL 35801-2031
(205) 536-3904

Charlotte VEC
227 Bennett Lane
Charlotte, NC 28213-6719
(704) 596-2168

Golden Empire Amateur Radio Society
P.O. Box 508
Chico, CA 95927-0508
(916) 342-1180

Great Lakes Amateur Radio Club VEC Inc.
P.O. Box 273
Glenview, IL 60025-0273
(708) 486-8019

Greater Los Angeles Amateur Radio Group
9737 Noble Avenue
Sepulveda, CA 91343-2403
(818) 892-2068
FAX (818) 892-9855

Jefferson Amateur Radio Club
P.O. Box 24368
New Orleans, LA 70184-4368

Koolau Amateur Radio Club
45-529 Nakuluai Street
Kaneohe, HI 96744-2224
(808) 235-4132

Laurel Amateur Radio Club, Inc.
P.O. Box 3039
Laurel, MD 20709-0039
(301) 572-5124 (0900-2100)
(301) 317-7819

The Milwaukee Radio Amateurs Club, Inc.
1737 N. 116th Street
Wauwatosa, WI 53226-3003
(414) 774-6999

Mountain Amateur Radio Club
P.O. Box 10
Burlington, WV 26710-0010
(304) 289-3576
(301) 724-0674

PHD Amateur Radio Association, Inc.
P.O. Box 11
Liberty, MO 64068-0011
(816) 781-7313

Sandarc-VEC
P.O. Box 2446
La Mesa, CA 91943-2446
(619) 465-3926

Sunnyvale VEC Amateur Radio Club
P.O. Box 60307
Sunnyvale, CA 94088-0307
(408) 255-9000 (24 hours)

Triad Emergency Amateur Radio Club
3504 Stonehurst Place
High Point, NC 27265-2106
(919) 841-7576

Western Carolina Amateur Radio
   Society VEC, Inc.
5833 Clinton Hw, Suite 203
Knoxville, TN 37912-2545
(615) 688-7771
FAX (615) 689-7062

W5YI-VEC*
P.O. Box 565101
Dallas, TX 75356-5101
(817) 461-6443
FAX (817) 548-9594

* This VEC regularly offers monthly exams in all parts of the country.

Write for graduation certificates to:
    Gordon West's Radio School
    2414 College Drive
    Costa Mesa, CA 92626

Please include:
  1. Certificate of completion
  2. One triple-stamped self-addressed
     envelope

AUTHORIZED FREQUENCY BANDS – AMATEUR SERVICE
(for U.S. Amateur Stations operating from ITU-Region 2–North and South America)

Meters	Novice	Technician[1,2]	Technician Plus[2]	General	Advanced	Extra Class
160				1800-2000 kHz/All	1800-2000 kHz/All	1800-2000 kHz/All
80	3675-3725 kHz/CW		3675-3725 kHz/CW	3525-3750 kHz/CW 3850-4000 kHz/Ph	3525-3750 kHz/CW 3775-4000 kHz/Ph	3500-4000 kHz/CW 3750-4000 kHz/Ph
40	7100-7150 KHz/CW		7100-7150 kHz/CW	7025-7150 kHz/CW 7225-7300 kHz/Ph	7025-7300 kHz/CW 7150-7300 kHz/Ph	7000-7300 kHz/CW 7150-7300 kHz/Ph
30				10.1-10.15 MHz/CW	10.1-10.15 MHz/CW	10.1-10.15 MHz/CW
20				14.025-14.15 MHz/CW 14.225-14.35 MHz/Ph	14.025-14.15 MHz/CW 14.175-14.35 MHz/Ph	14.0-14.35 MHz/CW 14.15-14.35 MHz/Ph
17				18.068-18.11 MHz/CW 18.11-18.168 MHz/Ph	18.068-18.11 MHz/CW 18.11-18.168 MHz/Ph	18.068-18.11 MHz/CW 18.11-18.168 MHz/Ph
15	21.1-21.2 MHz/CW		21.1-21.2 MHz/CW	21.025-21.2 MHz/CW 21.3-21.45 MHz/Ph	21.025-21.2 MHz/CW 21.225-21.45 MHz/Ph	21.0-21.45 MHz/CW 21.2-21.45 MHz/Ph
12				24.89-24.99 MHz/CW 24.93-24.99 MHz/Ph	24.89-24.99 MHz/CW 24.93-24.99 MHz/Ph	24.89-24.99 MHz/CW 24.93-24.99 MHz/Ph
10	28.1-28.5 MHz/CW 28.3-28.5 MHz/Ph		28.1-28.5 MHz/CW 28.3-28.5 MHz/Ph	28.0-29.7 MHz/CW 28.3-29.7 MHz/Ph	28.0-29.7 MHz/CW 28.3-29.7 MHz/Ph	28.0-29.7 MHz/CW 28.3-29.7 MHz/Ph
6	50-54 MHz/CW 50.1-54 MHz/Ph	50-54 MHz/CW 50.1-54 MHz/Ph	50-54 MHz/CW 50.1-54 MHz/Ph	50-54 MHz/CW 50.1-54 MHz/Ph	50-54 MHz/CW 50.1-54 MHz/Ph	50-54 MHz/CW 50.1-54 MHz/Ph
2	144-148 MHz/CW 144.1-148 MHz/All	144-148 MHz/CW 144.1-148 MHz/All	144-148 MHz/CW 144.1-148 MHz/All	144-148 MHz/CW 144.1-148 MHz/All	144-148 MHz/CW 144.1-148 MHz/All	144-148 MHz/CW 144.1-148 MHz/All
1.25	222.1-223.91 MHz/All	222-225 MHz/All	222-225 MHz/All	222-225 MHz/All	222-225 MHz/All	222-225 MHz/All
0.70	420-450 MHz/All	420-450 MHz/All	420-450 MHz/All	420-450 MHz/All	420-450 MHz/All	420-450 MHz/All
0.33	902-928 MHz/All	902-928 MHz/All	902-928 MHz/All	902-928 MHz/All	902-928 MHz/All	902-928 MHz/All
0.23	1270-1295 MHz/All	1240-1300 MHz/All	1240-1300 MHz/All	1240-1300 MHz/All	1240-1300 MHz/All	1240-1300 MHz/All

[1]No-Code License   [2]Effective 2/14/91

Note: Morse code (CW, A1A) may be used on any frequency allocated to the amateur service. Telephony emission (abbreviated Ph above) authorized on certain bands as indicated. Higher class licensees may use slow-scan television and facsimile emissions on the Phone bands; radio teletype/digital on the CW bands. All amateur modes and emissions are authorized above 144.1 MHz. In actual practice, the modes/emissions used are somewhat more complicated than shown above due to the existence of various band plans and "gentlemen's agreements" concerning where certain operations should take place.

SUMMARY OF FIGURES AND QUESTION POOL FORMULAS FOR NOVICE ELEMENT 2

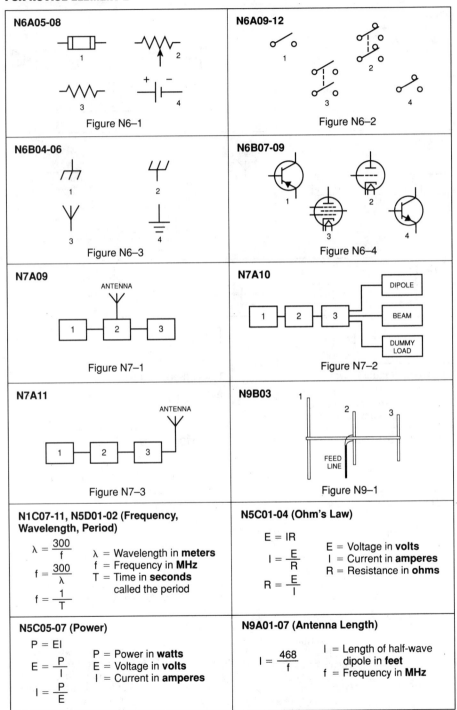

N6A05-08 Figure N6–1	**N6A09-12** Figure N6–2
N6B04-06 Figure N6–3	**N6B07-09** Figure N6–4
N7A09 Figure N7–1	**N7A10** Figure N7–2
N7A11 Figure N7–3	**N9B03** Figure N9–1

N1C07-11, N5D01-02 (Frequency, Wavelength, Period)

$$\lambda = \frac{300}{f}$$

$$f = \frac{300}{\lambda}$$

$$f = \frac{1}{T}$$

λ = Wavelength in **meters**
f = Frequency in **MHz**
T = Time in **seconds** called the period

N5C01-04 (Ohm's Law)

$$E = IR$$

$$I = \frac{E}{R}$$

$$R = \frac{E}{I}$$

E = Voltage in **volts**
I = Current in **amperes**
R = Resistance in **ohms**

N5C05-07 (Power)

$$P = EI$$

$$E = \frac{P}{I}$$

$$I = \frac{P}{E}$$

P = Power in **watts**
E = Voltage in **volts**
I = Current in **amperes**

N9A01-07 (Antenna Length)

$$I = \frac{468}{f}$$

I = Length of half-wave dipole in **feet**
f = Frequency in **MHz**

SUMMARY OF FIGURES AND QUESTION POOL FORMULAS
FOR TECHNICIAN ELEMENT 3A

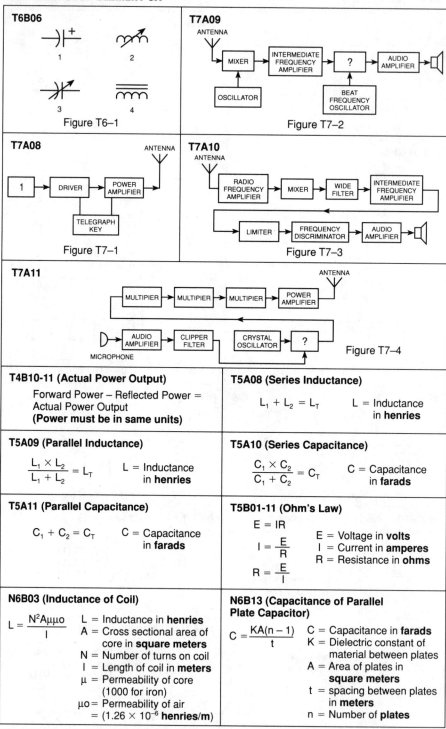

T6B06

Figure T6–1

T7A09

Figure T7–2

T7A08

Figure T7–1

T7A10

Figure T7–3

T7A11

Figure T7–4

T4B10-11 (Actual Power Output)

Forward Power − Reflected Power = Actual Power Output
(Power must be in same units)

T5A08 (Series Inductance)

$$L_1 + L_2 = L_T \qquad L = \text{Inductance in } \textbf{henries}$$

T5A09 (Parallel Inductance)

$$\frac{L_1 \times L_2}{L_1 + L_2} = L_T \qquad L = \text{Inductance in } \textbf{henries}$$

T5A10 (Series Capacitance)

$$\frac{C_1 \times C_2}{C_1 + C_2} = C_T \qquad C = \text{Capacitance in } \textbf{farads}$$

T5A11 (Parallel Capacitance)

$$C_1 + C_2 = C_T \qquad C = \text{Capacitance in } \textbf{farads}$$

T5B01-11 (Ohm's Law)

$$E = IR$$
$$I = \frac{E}{R}$$
$$R = \frac{E}{I}$$

E = Voltage in **volts**
I = Current in **amperes**
R = Resistance in **ohms**

N6B03 (Inductance of Coil)

$$L = \frac{N^2 A \mu \mu o}{I}$$

L = Inductance in **henries**
A = Cross sectional area of core in **square meters**
N = Number of turns on coil
I = Length of coil in **meters**
μ = Permeability of core (1000 for iron)
μo = Permeability of air = $(1.26 \times 10^{-6}$ **henries/m**)

N6B13 (Capacitance of Parallel Plate Capacitor)

$$C = \frac{KA(n-1)}{t}$$

C = Capacitance in **farads**
K = Dielectric constant of material between plates
A = Area of plates in **square meters**
t = spacing between plates in **meters**
n = Number of **plates**

ITU REGIONS

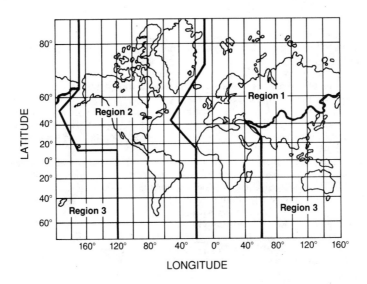

US CALL SIGNS

POPULAR Q SIGNALS

Given below are a number of Q signals whose meanings most often need to be expressed with brevity and clarity in amateur work. (Q abbreviations take the form of questions only when each is sent followed by a question mark.)

QRA   The name of my station is _____ .

QRG   Will you tell me my exact frequency (or that of _____)? Your exact frequency (or that of _____) is _____ kHz.

QRH   Does my frequency vary? Your frequency varies.

QRI   How is the tone of my transmission? The tone of your transmission is _____ (1. Good; 2. Variable; 3. Bad).

QRJ   Are you receiving me badly? I cannot receive you. Your signals are too weak.

QRK   What is the intelligibility of my signals (or those of _____)? The intelligibility of your signals (or those of _____) is _____ (1. Bad; 2. Poor; 3. Fair; 4. Good; 5. Excellent).

QRL   Are you busy? I am busy (or I am busy with _____). Please do not interfere.

QRM   Is my transmission being interfered with? Your transmission is being interfered with _____ (1. Nil; 2. Slightly; 3. Moderately; 4. Severely; 5. Extremely).

QRN   Are you troubled by static? I am troubled by static _____ (1-5 as under QRM).

QRO   Shall I increase power? Increase power.

QRP   Shall I decrease power? Decrease power.

QRQ   Shall I send faster? Send faster (_____ WPM).

QRS   Shall I send more slowly? Send more slowly (_____ WPM).

QRT   Shall I stop sending? Stop sending.

QRU   Have you anything for me? I have nothing for you.

QRV   Are you ready? I am ready.

QRW   Shall I inform _____ that you are calling on _____ kHz? Please inform _____ that I am calling on _____ kHz.

QRX   When will you call me again? I will call you again at _____ hours (on _____ kHz).

QRY   What is my turn? Your turn is numbered _____ .

QRZ   Who is calling me? You are being called by _____ (on _____ kHz).

QSA   What is the strength of my signals (or those of _____)? The strength of your signals (or those of _____) is _____ (1. Scarcely perceptible; 2. Weak; 3. Fairly good; 4. Good; 5. Very good).

QSB   Are my signals fading? Your signals are fading.

QSD   Is my keying defective? Your keying is defective.

QSG   Shall I send _____ messages at a time? Send _____ messages at a time.

QSK   Can you hear me between your signals and if so can I break in on your transmission? I can hear you between my signals; break in on my transmission.

QSL   Can you acknowledge receipt? I am acknowledging receipt.

QSM   Shall I repeat the last message which I sent you, or some previous message? Repeat the last message which you sent me [or message(s) number(s) _____].

QSN   Did you hear me (or _____) on _____ kHz? I heard you (or _____) on _____ kHz.

QSO   Can you communicate with _____ direct or by relay? I can communicate with _____ direct (or by relay through _____).

QSP   Will you relay to _____ ? I will relay to _____ .

QST   General call preceding a message addressed to all amateurs and ARRL members. This is in effect "CQ ARRL."

QSU   Shall I send or reply on this frequency (or on _____ kHz)?

QSW   Will you send on this frequency (or on _____ kHz)? I am going to send on this frequency (or on _____ kHz).

QSX   Will you listen to _____ on _____ kHz? I am listening to _____ on _____ kHz.

QSY   Shall I change to transmission on another frequency? Change to transmission on another frequency (or on _____ kHz).

QSZ   Shall I send each word or group more than once? Send each word or group twice (or _____ times).

QTA   Shall I cancel message number _____ ? Cancel message number _____ .

QTB Do you agree with my counting of words? I do not agree with your counting of words. I will repeat the first letter or digit of each word or group.
QTC How many messages have you to send? I have messages for you (or for _____).
QTH What is your location? My location is _____.
QTR What is the correct time? The time is _____.

Source: ARRL

COMMON CW ABBREVIATIONS

AA	All after	**NR**	Number
AB	All before	**NW**	Now; I resume transmission
ABT	About		
ADR	Address	**OB**	Old boy
AGN	Again	**OM**	Old man
ANT	Antenna	**OP-OPR**	Operator
BCI	Broadcast interference	**OT**	Old timer; old top
BK	Break; break me; break in	**PBL**	Preable
BN	All between; been	**PSE-PLS**	Please
B4	Before	**PWR**	Power
C	Yes	**PX**	Press
CFM	Confirm; I confirm	**R**	Received as transmitted; are
CK	Check		
CL	I am closing my station; call	**RCD**	Received
		REF	Refer to; referring to; reference
CLD-CLG	Called; calling	**RPT**	Repeat; I repeat
CUD	Could	**SED**	Said
CUL	See you later	**SEZ**	Says
CUM	Come	**SIG**	Signature; signal
CW	Continuous Wave	**SKED**	Schedule
DLD-DLVD	Delivered	**SRI**	Sorry
DX	Distance	**SVC**	Service; prefix to service message
FB	Fine business; excellent		
GA	Go ahead (or resume sending)	**TFC**	Traffic
		TMW	Tomorrow
GB	Good-by	**TNX**	Thanks
GBA	Give better address	**TU**	Thank you
GE	Good evening	**TVI**	Television interference
GG	Going	**TXT**	Text
GM	Good morning	**UR-URS**	Your; you're; yours
GN	Good night	**VFO-**	Variable-frequency oscillator
GND	Ground		
GUD	Good	**VY**	Very
HI	The telegraphic laugh; high	**WA**	Word after
		WB	Word before
HR	Here; hear	**WD-WDS**	Word; words
HV	Have	**WKD-WKG**	Worked; working
HW	How	**WL**	Well; will
LID	A poor operator	**WUD**	Would
MILS	Milliamperes	**WX**	Weather
MSG	Message; prefix to radiogram	**XMTR**	Transmitter
		XTAL	Crystal
N	No	**XYL**	Wife
ND	Nothing doing	**YL**	Young lady
NIL	Nothing; I have nothing for you	**73**	Best regards
		88	Love and kisses

LIST OF COUNTRIES PERMITTING THIRD-PARTY TRAFFIC

V2	Antigua & Barbuda	C5	The Gambia	DU	Philippines
LU	Argentina	9G	Ghana	VR6	Pitcairn Island*
VK	Australia	J3	Grenada	V4	St Christopher (St. Kitts)
V3	Belize	8R	Guyana		and Nevis
CP	Bolivia	HH	Haiti	J6	St. Lucia
PY	Brazil	HR	Honduras	J8	St. Vincent
VE	Canada	4X	Israel	9L	Sierra Leone
CE	Chile	6Y	Jamaica	3D6	Swaziland
HK	Colombia	JY	Jordan	9Y	Trinidad & Tobago
D6	Comoros	EL	Liberia	GB	United Kingdom**
TI	Costa Rica	XE	Mexico	CX	Uruguay
CO	Cuba	V6	Micronesia	YV	Venezuela
J7	Dominica	YN	Nicaragua	4U1ITU	ITU, Geneva
HI	Dominican Rep.	HP	Panama	4U1VIC	VIC, Geneva
HC	Ecuador	ZP	Paraguay		
YS	El Salvador	OA	Peru		

* Informal Temporary
** Limited to special-event stations with callsign prefix GB; GB3 excluded   3/93

COUNTRIES HOLDING U.S. RECIPROCAL AGREEMENTS

Antigua	Germany (Fed. Rep.)	Netherlands Ant.
Argentina	Greece	New Zealand
Australia	Grenada	Nicaragua
Austria	Guatemala	Norway
Bahamas	Guyana	Panama
Barbados	Haita	Paraguay
Belgium	Honduras	Papua New Guinea
Belize	Hong Kong	Peru
Bolivia	Iceland	Phillippineas
Botswana	India	Portugal
Brazil	Indonesia	St. Lucia
Canada	Ireland	Sierra Leone
Chile	Israel	Solomon Islands
Colombia	Italy	South Africa
Costa Rica	Jamaica	Spain
Cyprus	Japan	Suriname
Denmark	Jordan	Sweden
Dominica (Comwth)	Kiribati	Switzerland
Dominican Rep.	Kuwait	Trinidad
Ecuador	Liberia	Tuvalu
El Salvador	Luxembourg	United Kingdom*
Fiji	Mexico	Uruguay
Finland	Monaco	Venezuela
France*	Netherlands	Yugoslavia

* Includes many British and French countries and territories

3/93

EMISSION TYPE DESIGNATORS – AMATEUR RADIO SERVICE

First Symbol – Modulation system

A=Double sideband AM, C=Vestigial sideband AM, D=Amplitude/angle modulated, F=Frequency modulation, G=Phase modulation, H=Single sideband/full carrier, J=Single sideband/suppressed carrier, K=AM pulse, L=Pulse modulated in width/duration, M=Pulse modulated in position/phase, P=Unmodulated pulses, Q=Angle modulated during pulse, V=Combination of pulse emissions, W=Other types of pulses, R=Single sideband/reduced or variable level carrier.

Second Symbol – Nature of signal modulating carrier

0=No modulation, 1=Digital data without modulated subcarrier, 2=Digital data on modulated subcarrier, 3=Analog modulated, 7=Two or more channels of digital data, 8=Two or more channels of analog data, 9=Combination of analog and digital information, X=Other

Third Symbol – Information to be conveyed

A=Manually received telegraphy, B=Automatically received telegraphy, C=Facsimile (FAX), D=Digital information, E=Voice telephony, F=Video/television, N=No information, W=Combination of these, X=Other.

1. **CW** - International Morse code telegraphy emissions having designators with A, C, H, J or R as the first symbol, 1 as the second symbol, A or B as the third symbol.

2. **DATA** - Telemetry, telecommand and computer communications emissions having designators with A, C, D, F, G, H, J or R as the first symbol: 1 as the second symbol; D as the third symbol; and also emission J2D. Only a digital code of a type specifically authorized in the §Part 97.3 rules may be transmitted.

3. **IMAGE** - Facsimile and television emissions having designators with A, C, D, F, G, H, J or R as the first symbol; 1, 2 or 3 as the second symbol; C or F as the third symbol; and also emissions having B as the first symbol; 7, 8 or 9 as the second symbol; W as the third symbol.

4. **MCW (Modulated carrier wave)** - Tone-modulated international Morse code telegraphy emissions having designators with A, C, D, F, G, H or R as the first symbol; 2 as the second symbol; A or B as the third symbol.

5. **PHONE** - Speech and other sound emissions having designators with A, C, D, F, G, H, J or R as the first symbol; 1, 2 or 3 as the second symbol; E as the third symbol. Also speech emissions having B as the first symbol; 7, 8 or 9 as the second symbol; E as the third symbol. MCW for the purpose of performing the station identification procedure, or for providing telegraphy practice interspersed with speech, or incidental tones for the purpose of selective calling, or alerting or to control the level of a demodulated signal may also be considered phone.

6. **PULSE** - Emissions having designators with K, L, M, P, Q, V or W as the first symbol; , 1, 2, 3, 7, 8, 9 or X as the second symbol; A, B, C, D, E, F, N, W or X as the third symbol.

7. **RTTY (Radioteletype)** - Narrow-band, direct-printing telegraphy emissions having designators with A, C, D, F, G, H, J or R as the first symbol; 1 as the second symbol; B as the third symbol; and also emission J2B. Only a digital code of a type specifically authorized in the §Part 97.3 rules may be transmitted.

8. **SS (Spread Spectrum)** - Emissions using bandwidth-expansion modulation emissions having designators with A, C, D, F, G, H, J or R as the first symbol; X as the second symbol; X as the third symbol. Only a SS emission of a type specifically authorized in §Part 97.3 rules may be transmitted.

9. **TEST** - Emissions containing no information having the designators with N as the third symbol. Test does not include pulse emissions with no information or modulation unless pulse emissions are also authorized in the frequency band.

Attach the original license or photocopy here

FEDERAL COMMUNICATIONS COMMISSION
GETTYSBURG, PA 17325-7245

Approved OMB 3060-0003
Expires 02/28/95
See instructions for information regarding public burden estimate

APPLICATION FOR AMATEUR RADIO STATION/OPERATOR LICENSE

SECTION II—EXAMINATION INFORMATION

CERTIFICATION BY ALL VEs

I CERTIFY THAT I have complied with the Administering VE requirements stated in Part 97 of the Commission's Rules; THAT I have administered to the applicant all required elements for an amateur radio operator examination in accordance with Part 97 of the Commission's Rules; THAT I have indicated in the Administering VEs Report the examination element(s) the applicant passed; THAT I have examined documents held by the applicant and I have indicated in the Administering VEs Report the examination element for which the applicant is given examination credit in accordance with Part 97 of the Commission's Rules.

SECTION II-A FOR NOVICE OPERATOR EXAMINATION ONLY. To be completed by the Administering VE's after completing the Administering VEs Report on the other side of this form.

SECTION II-B FOR TECHNICIAN, GENERAL, ADVANCED, OR AMATEUR EXTRA OPERATOR EXAMINATION ONLY. To be completed by the Administering VE's after completing the Administering VEs Report on the other side of this form.

PHYSICIAN'S CERTIFICATION OF DISABILITY

WILLFUL FALSE STATEMENTS ARE PUNISHABLE BY FINE AND IMPRISONMENT, U.S. CODE TITLE 18, SECTION 1001.

FCC Form 610, March 1992

b. Back

ADMINISTERING VE's REPORT

EXAMINATION ELEMENTS

SECTION I

CERTIFICATION

1. CURRENT MAILING ADDRESS: P.O. Box 2468

CITY: FREESVILLE   STATE: CA   ZIP CODE: 91234

LAST NAME: HAMMER   M.I.: F   CURRENT FIRST NAME: SUSIE

6. DATE OF BIRTH: 06-22-42

CURRENT STATION LOCATION: 12345 PEAR TREE LANE   CITY: FREESVILLE   STATE: CA

SIGNATURE OF APPLICANT: Susie F. Hammer   14. DATE SIGNED: 6/8/92

FCC Form 610, March 1992

a. Front

(Filled-in area for applicant, shaded area for VEs)

Form 610 Novice and Technician Class License Application Form

* Effective March 1, 1993, the FCC will ignore information in items 2H and 8. These items will be eliminated in new versions of the form.

228

Glossary

Amateur communication: Non-commercial radio communication by or among amateur stations solely with a personal aim and without personal or business interest.

Amateur operator/primary station license: An instrument of authorization issued by the Federal Communications Commission comprised of a station license, and also incorporating an operator license indicating the class of privileges.

Amateur operator: A person holding a valid license to operate an amateur station issued by the Federal Communications Commission. Amateur operators are frequently referred to as ham operators.

Amateur Radio services: The amateur service, the amateur-satellite service and the radio amateur civil emergency service.

Amateur-satellite service: A radiocommunication service using stations on Earth satellites for the same purpose as those of the amateur service.

Amateur service: A radiocommunication service for the purpose of self-training, intercommunication and technical investigations carried out by amateurs; that is, duly authorized persons interested in radio technique solely with a personal aim and without pecuniary interest.

Amateur station: A station licensed in the amateur service embracing necessary apparatus at a particular location used for amateur communication.

AMSAT: Radio Amateur Satellite Corporation, a non-profit scientific organization. (P.O. Box #27, Washington, DC 20044)

ARES: Amateur Radio Emergency Service — the emergency division of the American Radio Relay League. Also see RACES

ARRL: American Radio Relay League, national organization of U.S. Amateur Radio operators. (225 Main Street, Newington, CT 06111)

Audio Frequency (AF): The range of frequencies that can be heard by the human ear, generally 20 hertz to 20 kilohertz.

Automatic control: The use of devices and procedures for station control without the control operator being present at the control point when the station is transmitting.

Automatic Volume Control (AVC): A circuit that continually maintains a constant audio output volume in spite of deviations in input signal strength.

Beam or Yagi antenna: An antenna array that receives or transmits RF energy in a particular direction. Usually rotatable.

Block diagram: A simplified outline of an electronic system where circuits or components are shown as boxes.

Broadcasting: Information or programming transmitted by radio means intended for the general public.

Business communications: Any transmission or communication the purpose of which is to facilitate the regular business or commercial affairs of any party. Business communications are prohibited in the amateur service.

Call Book: A published list of all licensed amateur operators available in North American and Foreign editions.

Call sign assignment: The FCC systematically assigns each amateur station their primary call sign. The FCC will not grant a request for a specific call sign.

Certificate of Successful Completion of Examination (CSCE): A certificate of successful completion allowing examination credit for 365 days. Both written and code credit can be authorized.

Coaxial cable, Coax: A concentric two-conductor cable in which one conductor surrounds the other, separated by an insulator.

Control operator: An amateur operator designated by the licensee of an amateur station to be responsible for the station transmissions.

Coordinated repeater station: An amateur repeater station for which the transmitting and receiving frequencies have been recommended by the recognized repeater coordinator.

Coordinated Universal Time (UTC): Sometimes referred to as Greenwich Mean Time, UCT or Zulu time. The time at the zero-degree (0°) Meridian which passes through Greenwich, England. A universal time among all amateur operators.

Crystal: A quartz or similar material which has been ground to produce natural vibrations of a specific frequency. Quartz crystals produce a high degree of frequency stability in radio transmitters.

CW: Continuous wave, another term for the International Morse code.

Dipole antenna: The most common wire antenna. Length is equal to one-half of the wavelength. Fed by coaxial cable.

Dummy antenna: A device or resistor which serves as a transmitter's antenna without radiating radio waves. Generally used to tune up a radio transmitter.

Duplexer: A device that allows a single antenna to be simultaneously used for both reception and transmission.

Effective Radiated Power (ERP): The product of the transmitter (peak envelope) power, expressed in watts, delivered to the antenna, and the relative gain of an antenna over that of a half wave dipole antenna.

Emergency communication: Any amateur communication directly relating to the immediate safety of life of individuals or the immediate protection of property.

Examination Credit Certificate: (See Certificate of Successful Completion of Examination)

Examination Element: Novices must pass Element 1(A) 5-wpm code test and an Element 2 written theory examination. Technician no-code must pass Element 2 and 3A written theory examinations.

FCC Form 610: The amateur service application form for an amateur operator/primary station license. It is used to apply for a new amateur license or to renew or modify an existing license.

Federal Communications Commission (FCC): A board of five Commissioners, appointed by the President, having the power to regulate wire and radio telecommunications in the United States.

Feedline Transmission line: A system of conductors that connects an antenna to a receiver or transmitter.

Field Day: Annual activity sponsored by the ARRL to demonstrate emergency preparedness of amateur operators.

Filter: A device used to block or reduce alternating currents or signals at certain frequencies while allowing others to pass unimpeded.

Frequency: The number of cycles of alternating current in one second.

Frequency coordinator: An individual or organization recognized by amateur operators eligible to engage in repeater operation which recommends frequencies and other operating and/or technical parameters for amateur repeater operation in order to avoid or minimize potential interferences.

Frequency Modulation (FM): A method of varying a radio carrier wave by causing its frequency to vary in accordance with the information to be conveyed.

Frequency privileges: The transmitting frequency bands available to the various classes of amateur operators. The Novice and Technician privileges are listed in Part 97.301 of the FCC rules.

Ground: A connection, accidental or intentional, between a device or circuit and the earth or some common body and the earth or some common body serving as the earth.

Ground wave: A radio wave that is propagated near or at the earth's surface.

Handi-Ham system: Amateur organization dedicated to assisting handicapped amateur operators. (3915 Golden Valley Road, Golden Valley, MN 55422)

Harmful interference: Interference which seriously degrades, obstructs or repeatedly interrupts the operation of a radio communication service.

Harmonic: A radio wave that is a multiple of the fundamental frequency. The second harmonic is twice the fundamental frequency, the third harmonic, three times, etc.

Hertz: One complete alternating cycle per second. Named after Heinrich R. Hertz, a German physicist. The number of hertz is the frequency of the audio or radio wave.

High Frequency (HF): The band of frequencies that lie between 3 and 30 Megahertz. It is from these frequencies that radio waves are returned to earth from the ionosphere.

High-Pass filter: A device that allows passage of high frequency signals but attenuates the lower frequencies. When installed on a television set, a high-pass filter allows TV frequencies to pass while blocking lower frequency amateur signals.

Ionosphere: Outer limits of atmosphere from which HF amateur communications signals are returned to earth.

Jamming: The intentional malicious interference with another radio signal.

Key clicks, Chirps: Defective keying of a telegraphy signal sounding like tapping or high varying pitches.

Lid: Amateur slang term for poor radio operator.

Linear amplifier: A device that accurately reproduces a radio wave in magnified form.

Long wire: A horizontal wire antenna that is one wavelength or longer in length.

Low-Pass filter: Device connected to worldwide transmitters that inhibits passage of higher frequencies that cause television interference but does not affect amateur transmissions.

Machine: A ham slang word for an automatic repeater station.

Malicious interference: Willful, intentional jamming of radio transmissions.

MARS: The Military Affiliate Radio System. An organization that coordinates the activities of amateur communications with military radio communications.

Maximum authorized transmitting power: Amateur stations must use no more than the maximum transmitter power necessary to carry out the desired communications. The maximum P.E.P. output power levels authorized Novices are 200 watts in the 80-, 40-, 15- and 10-meter bands, 25 watts in the 222-MHz band, and 5 watts in the 1270-MHz bands.

Maximum usable frequency: The highest frequency that will be returned to earth from the ionosphere.

Medium frequency (MF): The band of frequencies that lies between 300 and 3,000 kHz (3 MHz).

Mobile operation: Radio communications conducted while in motion or during halts at unspecified locations.

Mode: Type of transmission such as voice, teletype, code, television, facsimile.

Modulate: To vary the amplitude, frequency or phase of a radio frequency wave in accordance with the information to be conveyed.

Morse code (see CW): The International Morse code, A1A emission. Interrupted continuous wave communications conducted using a dot-dash code for letters, numbers and operating procedure signs.

No-Code Technician operator: An amateur radio operator who has successfully passed Element 2 and 3A. This operator has not passed Element 1A, the 5-wpm code test.

Novice operator: An FCC licensed entry-level amateur operator in the amateur service. Novices may operate a transmitter in the following meter wavelength bands: 80, 40, 15, 10, 1.25 and 0.23.

Ohm's law: The basic electrical law explaining the relationship between voltage, current and resistance. The current I in a circuit is equal to the voltage E divided by the resistance R, or I = E/R.

OSCAR: Stands for "Orbiting Satellite Carrying Amateur Radio," the name given to a series of satellites designed and built by amateur operators of several nations.

Oscillator: A device for generating oscillations or vibrations of an audio or radio frequency signal.

Packet radio: A digital method of communicating computer-to-computer. A terminal-node controller makes up the packet of data and directs it to another packet station.

Peak Envelope Power (PEP): 1. The power during one radio frequency cycle at the crest of the modulation envelope, taken under normal operating conditions. 2. The maximum power that can be obtained from a transmitter.

Phone patch: Interconnection of amateur service to the public switched telephone network, and operated by the control operator of the station.

Power supply: A device or circuit that provides the appropriate voltage and current to another device or circuit.

Propagation: The travel of electromagnetic waves or sound waves through a medium.

Q-signals: International three-letter abbreviations beginning with the letter Q used primarily to convey information using the Morse code.

QSL Bureau: An office that bulk processes QSL (radio confirmation) cards for (or from) foreign amateur operators as a postage saving mechanism.

RACES (radio amateur civil emergency service): A radio service using amateur stations for civil defense communications during periods of local, regional, or national civil emergencies.

Radiation: Electromagnetic energy, such as radio waves, traveling forth into space from a transmitter.

Radio Frequency (RF): The range of frequencies over 20 kilohertz that can be propagated through space.

Radio wave: A combination of electric and magnetic fields varying at a radio frequency and traveling through space at the speed of light.

Repeater operation: Automatic amateur stations that retransmit the signals of other amateur stations. Novices and Technicians may operate through amateur repeaters.

RST Report: A telegraphy signal report system of Readability, Strength and Tone.

S-meter: A voltmeter calibrated from 0 to 9 that indicates the relative signal strength of an incoming signal at a radio receiver.

Selectivity: The ability of a circuit (or radio receiver) to separate the desired signal from those not wanted.

Sensitivity: The ability of a circuit (or radio receiver) to detect a specified input signal.

Short circuit: An unintended low resistance connection across a voltage source resulting in high current and possible damage.

Shortwave: The high frequencies that lie between 3 and 30 Megahertz that are propagated long distances.

Single-Sideband (SSB): A method of radio transmission in which the RF carrier and one of the sidebands is suppressed and all of the information is carried in the one remaining sideband.

Skip wave, Skip zone: A radio wave reflected back to earth. The distance between the radio transmitter and the site of a radio wave's return to earth.

Sky wave: A radio wave that is refracted back to earth in much the same way that a stone thrown across water skips out. Sometimes called an ionospheric wave.

Spectrum: A series of radiated energies arranged in order of wavelength. The radio spectrum extends from 20 kilohertz upward.

Spurious Emissions: Unwanted radio frequency signals emitted from a transmitter that sometimes causes interference.

Station license, location: No transmitting station shall be operated in the amateur service without being licensed by the Federal Communications Commission. Each amateur station shall have one land location, the address of which appears in the station license.

Sunspot Cycle: An 11-year cycle of solar disturbances which greatly affects radio wave propagation.

Technician: A no-code amateur operator who has all privileges from 6 meters on up to shorter wavelengths, but *no* privileges on 10, 15, 40, or 80 meter wavelength bands.

Technician Plus: An amateur operator who has passed a 5-wpm code test in addition to Technician Class requirements.

Telegraphy: Telegraphy is communications transmission and reception using CW, International Morse Code.

Telephony: Telephony is communications transmission and reception in the voice mode.

Telecommunications: The electrical conversion, switching, transmission and control of audio signals by wire or radio. Also includes video and data communications.

Temporary operating authority: Authority to operate your amateur station while awaiting arrival of an upgraded license. New Novice operators are not granted temporary operating authority and must await receipt of their new license and call sign.

Terrestrial station location: Any location of a radio station on the surface of the earth including the sea.

Third-party traffic: Amateur communication by or under the supervision of the control operator at an amateur station to another amateur station on behalf of others.

Transceiver: A combination radio transmitter and receiver.

Transmatch: An antenna tuner used to match the impedance of the transmitter output to the transmission line of an antenna.

Transmitter: Equipment used to generate radio waves. Most commonly, this radio carrier signal is amplitude varied or frequency varied (modulated) with information and radiated into space.

Transmitter power: The average peak envelope power (output) present at the antenna terminals of the transmitter. The term "transmitted" includes any external radio frequency power amplifier which may be used.

Ultra High Frequency (UHF): Ultra high frequency radio waves that are in the range of 300 to 3,000 MHz.

Upper Sideband (USB): The proper operating mode for sideband transmissions made in the new Novice 10-meter voice band. Amateurs generally operate USB at 20 meters and higher frequencies; lower sideband (LSB) at 40 meters and lower frequencies.

Very High Frequency (VHF): Very high frequency radio waves that are in the range of 30 to 300 MHz.

Volunteer Examiner: An amateur operator of at least a General Class level who administers or prepares amateur operator license examinations. A VE must be at least 18 years old and not related to the applicant.

Volunteer Examiner Coordinator (VEC): A member of an organization which has entered into an agreement with the FCC to coordinate the efforts of volunteer examiners in preparing and administering examinations for amateur operator licenses.

Index

☞ **Dear Reader:** *We'd like your views on the books we publish.*

PROMPT® Publications, an imprint of Howard W. Sams & Company, is dedicated to bringing you timely and authoritative documentation and information you can use.

You can help us in our continuing effort to meet your information needs. Please take a few moments to answer the questions below. Your answers will help us serve you better in the future.

1. What is the title of the book you purchased? _____

2. Where do you usually buy books? _____

3. Where did you buy this book? _____

4. Was the information useful? _____

5. What did you like most about the book? _____

6. What did you like least? _____

7. Is there any other information you'd like included? _____

8. In what subject areas would you like us to publish more books?

    (Please check the boxes next to your fields of interest.)

    ❏ Amateur Radio   ❏ Computer Software

    ❏ Antique Radio and TV   ❏ Electronics Concepts Theory

    ❏ Audio Equipment Repair   ❏ Electronics Projects/Hobbies

    ❏ Camcorder Repair   ❏ Home Appliance Repair

    ❏ Computer Hardware   ❏ TV Repair

    ❏ Computer Programming   ❏ VCR Repair

9. Are there other subjects not covered in the checklist that you'd like to see books about?

    _____

10. Comments _____

    _____

    _____

Name _____

Address _____

City_____ State/Zip _____

Occupation_____ Daytime Phone _____

Thanks for helping us make our books better for all of our readers. Please drop this postage-paid card in the nearest mailbox.

For more information about PROMPT®Publications,
see your authorized Sams PHOTOFACT®distributor.
Or call 1-800-428-7267 for the name of your nearest PROMPT®Publications
distributor.

Imprint of Howard W. Sams & Company

2647 Waterfront Parkway East Drive,

Indianapolis, IN 46214-2041

BUSINESS REPLY MAIL

FIRST CLASS MAIL   PERMIT NO. 1317   INDIANAPOLIS IN

POSTAGE WILL BE PAID BY ADDRESSEE

HOWARD W. SAMS & COMPANY

2647 WATERFRONT PKY EAST DR
INDIANAPOLIS IN 46209-1418